D. WALTER STUART

The Complete Guide to the 2024 Solar Eclipse

Essential Tips to Enhance Your Eclipse Experience including Historical, Scientific, Travel and Photographic Information

Copyright © 2024 by D. Walter Stuart

All rights reserved. No part of this publication may be reproduced, stored or transmitted in any form or by any means, electronic, mechanical, photocopying, recording, scanning, or otherwise without written permission from the publisher. It is illegal to copy this book, post it to a website, or distribute it by any other means without permission.

D. Walter Stuart asserts the moral right to be identified as the author of this work.

D. Walter Stuart has no responsibility for the persistence or accuracy of URLs for external or third-party Internet Websites referred to in this publication and does not guarantee that any content on such Websites is, or will remain, accurate or appropriate.

Designations used by companies to distinguish their products are often claimed as trademarks. All brand names and product names used in this book and on its cover are trade names, service marks, trademarks and registered trademarks of their respective owners. The publishers and the book are not associated with any product or vendor mentioned in this book. None of the companies referenced within the book have endorsed the book.

First edition

This book was professionally typeset on Reedsy.
Find out more at reedsy.com

Contents

Introduction	1
Chapter 1: The Science Behind the Eclipse	5
Chapter 2: Historical Perspectives on Eclipses	13
Chapter 3: Finding Your Perfect Spot	23
Chapter 4: The Art (and Safety) of Eclipse Viewing	31
Chapter 5: Eclipse Photography for Beginners	43
Chapter 6: Creative Expressions of the Eclipse	51
Chapter 7: Environmental and Wildlife Responses to Eclipses	57
Chapter 8: The Philosophical and Existential Dimensions of...	65
Chapter 9: Looking Ahead – Future Eclipses and Continuing...	72
Chapter 10: Conclusion	78
Chapter 11: Resources	79
Share the joy — Leave a Review!	82

Introduction

"A solar eclipse is a cosmic billiard shot—the Sun, Moon, and Earth lining up to reveal the universe in motion." Carl Sagan

1. The Magic of Eclipses:

Solar eclipses have captivated human imagination for centuries, often seen as ominous signs or celestial warnings. In ancient China, eclipses were thought to be caused by mythical creatures like dragons attempting to devour the sun or moon, leading to rituals involving noise and shooting arrows to scare them off. The Vikings similarly attributed eclipses to mythical dogs chasing celestial bodies, suggesting diverse interpretations of eclipses across many cultures. Greek scientists were among the first to study solar eclipses scientifically, with Thales of Miletus credited with possibly the earliest recorded prediction. Ancient scientists developed mathematical methods to predict eclipses, which were essential for maintaining calendars and planning significant events like harvests or ceremonies.

In medieval times, the study of eclipses spurred advancements in trigonometry and astronomy. Astronomers like Al-Biruni and Ibn al-Haytham laid the foundation for future scientific progress. The Renaissance era brought enhanced accuracy to eclipse predictions, building on the mathematical models of astronomers like Ptolemy. Nicolaus Copernicus' heliocentric model revolutionized the general understanding of eclipse mechanics, while Johannes Kepler's laws of planetary motion significantly improved eclipse predictions. Galileo Galilei's telescopic observations of the moon's surface further enriched knowledge about the moon's role in eclipses, marking significant strides in the field.

Solar eclipses have contributed much to scientific understanding in the modern period. The 1919 complete solar eclipse was one significant example of the validation of Einstein's general relativity theory. Studies of eclipses were also conducted on the sun's corona.

We currently have a great deal of knowledge about solar eclipses and can accurately predict their occurrence. Due to their beauty and the remarkable ways they allow people to watch the sun, they continue to pique the interest of both amateur and professional astronomers.

2. Why the 2024 Eclipse is Special:

The 2024 total solar eclipse, expected to occur on April 8, 2024, has several unique and noteworthy characteristics:

The first total solar eclipse in the U.S. since 2017: The 2024 eclipse will be the first total solar eclipse visible in the United States since the widely-viewed August 21, 2017 event, thus making it particularly anticipated, especially for those who experienced or missed the 2017 eclipse.

Intersection with the 2017 Path of Totality: Interestingly, the path of the 2024 eclipse will intersect with the path of the 2017 eclipse. The intersection point, near Carbondale, Illinois, will experience totality in both eclipses, making it a unique spot for eclipse enthusiasts.

INTRODUCTION

Opportunity for Scientific Observations: Each eclipse offers unique opportunities for scientific study, particularly of the sun's corona. The 2024 eclipse will provide another valuable chance for astronomers and solar scientists to gather data and conduct experiments.

Path of Totality: The path of totality is the narrow track across the Earth's surface where the Moon completely covers the Sun during a total solar eclipse. Outside this path, only a partial eclipse is visible.

The 2024 eclipse will have a relatively wide path of totality, up to 115 miles in some areas. This path of totality will cross North America, starting in Mexico, crossing through the United States, and ending in Eastern Canada. This extensive path means the eclipse will be accessible to millions of people, including those in major cities. The 2017 U.S. eclipse, which traveled from the Pacific Northwest to the Southeastern U.S., or the 1999 eclipse in Europe, which passed across central Europe and into the Middle East, would have been viewable by fewer people.

The eclipse will have several phases:

- A partial eclipse begins when the Moon starts to cover the Sun. This phase is visible over a much larger area than the total phase.
- The total eclipse phase starts when the Moon completely covers the Sun. Observers in the path of totality will experience darkness, cooler temperatures, and the appearance of stars and planets.
- A maximum eclipse occurs at the peak moment when the Sun is entirely covered.
- The end of totality is when the Sun reappears, marking the end of the total phase.
- During the final partial phase, the Moon moves away from the Sun, gradually uncovering it until the eclipse ends.

The entire eclipse, from the start of the partial phase to the end, will last several hours, but the total phase at any given location is much shorter. Longer totality offers a more extended experience of the eclipse's full effects, such as the dramatic drop in temperature and the visibility of stars and planets.

High Public Interest and Educational Opportunities: Given the increased interest in astronomy and science, this eclipse is expected to be a major event for educational outreach, with numerous programs and activities planned around it.

Challenges in Viewing Due to Weather: The timing and path of the eclipse mean that weather could be a significant factor in viewing conditions. Early April can be unpredictable in terms of weather, especially in the northern parts of the path.

Tourism and Economic Impact: Like the 2017 eclipse, the 2024 event is expected to have a significant economic impact, drawing tourists to the path of totality and boosting local economies. A unique aspect of the 2024 eclipse is its intersection with the path of the 2017 eclipse, particularly around Carbondale, Illinois. This rare intersection creates a point of heightened interest for eclipse enthusiasts and scientists.

Frequency and Historical Significance: Total solar eclipses occur approximately every 18 months somewhere on Earth, but it's rare for them to pass over densely populated areas. The 2024 eclipse is significant as it follows the 2017 U.S. eclipse closely in historical terms, offering another viewing opportunity for those in North America.

In summary, while each total solar eclipse is a unique celestial event, the 2024 eclipse stands out for its path across a heavily populated region, the length of its totality, and its proximity in time and intersection with the path of the 2017 U.S. eclipse. This combination of factors is expected to make the 2024 eclipse a particularly notable event both for the general public and the scientific community.

Chapter 1: The Science Behind the Eclipse

"Bailey's Beads occur as the moon makes its final move over the sun, creating a dazzling string of luminous beads around the darkened moon, a fleeting yet unforgettable spectacle."

1. What Happens During an Eclipse?

When the Moon passes in front of the Earth and the Sun, it temporarily blocks the Sun's light, and a solar eclipse takes place. This creates the Sun's corona, or outer atmosphere, visible as a luminous halo surrounding the Moon. (See image below.) When the Moon partially obscures the Sun during a partial solar eclipse, the Sun takes on the shape of a crescent. A unique and breathtaking spectacle, solar eclipses are compelling celestial occurrences that happen when the Sun, Moon, and Earth align in a particular way.

Some fundamental astronomy ideas are necessary to comprehend how the Sun, Moon, and Earth line up during a solar eclipse:

Relative Sizes and Distances: The apparent sizes of the Sun and Moon in the sky during solar eclipses are among their most remarkable features. In addition to having a diameter around 400 times greater than that of the Moon, the Sun is also 400 times farther away from Earth. Because of this coincidence, it appears as though the Sun and Moon are about the same size in the sky.

A total solar eclipse is a rare and breathtaking celestial phenomenon made possible by the Moon's ability to cover the Sun totally due to their similar apparent sizes. The observed total solar eclipses would not happen if the Moon were appreciably bigger or smaller in the sky. This coincidental sizing adds to the beauty and rarity of total solar eclipses and is a unique characteristic of the Earth, Moon, and Sun systems.

Shadow Play: During a solar eclipse, the Moon casts a shadow on Earth. This shadow consists of two parts:

Umbra is the fully shaded inner region of a shadow where the Moon completely blocks the Sun. Observers located in the umbra will experience a total solar eclipse. The umbra is a relatively narrow path across the Earth's surface.

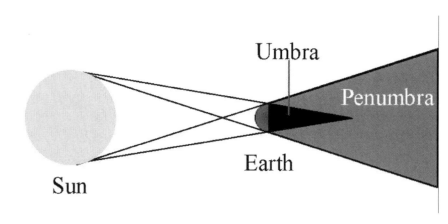

Penumbra: This is the partially shaded outer region of the shadow. Observers in this area see only a partial solar eclipse, where the Moon partially covers the Sun. The penumbral shadow covers a much larger area than the umbra but offers a less dramatic experience than the total eclipse observed in the umbra.

CHAPTER 1: THE SCIENCE BEHIND THE ECLIPSE

2. Path of Totality:

The path of the umbra across Earth's surface is known as the path of totality. The path of totality is essentially the route or track where viewers on Earth can observe the total solar eclipse. During a total solar eclipse, the Moon moves between the Earth and the Sun, completely covering the Sun for a brief period, thus casting a shadow on the Earth.

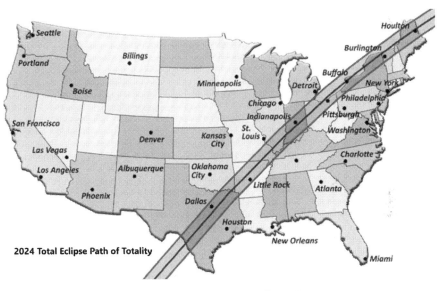

2024 Total Eclipse Path of Totality

For the April 8, 2024 eclipse, the path of totality will be a strip across North America, passing over Mexico, the United States, and Canada. The total solar eclipse will begin over the South Pacific Ocean. Weather permitting, the first location in continental North America to experience totality is Mexico's Pacific coast at around 11:07 a.m. PDT. The path of the eclipse continues from Mexico, entering the United States in Texas, and traveling through Oklahoma, Arkansas, Missouri, Illinois, Kentucky, Indiana, Ohio, Pennsylvania, New

York, Vermont, New Hampshire, and Maine. Small parts of Tennessee and Michigan will also experience the total solar eclipse. The eclipse will enter Canada in Southern Ontario and continue through Quebec, New Brunswick, Prince Edward Island, and Cape Breton. The eclipse will exit continental North America on the Atlantic coast of Newfoundland, Canada, at 5:16 p.m. NDT.

Cities directly in this path will experience the total eclipse, while areas outside this path will only see a partial eclipse. The closer you are to the center of the path, the longer the total eclipse will last. This rare event, and being in the path of totality, offers a unique and spectacular view of the eclipse.

These aspects of solar eclipses highlight the delicate balance and precision in celestial mechanics that allow us to witness these spectacular events. They also underscore the uniqueness of our place in the solar system, where such phenomena can be observed in dramatic fashion.

3. Types of Eclipses:

The specific alignment of the Sun, Moon, and Earth determines the type of eclipse. A total solar eclipse is when the Moon lines up precisely with the Sun in such a way that it completely obscures the Sun from Earth's perspective. We refer to this heavenly alignment as "syzygy." The sky darkens during a total solar eclipse, allowing onlookers to see the Sun's corona—its outermost layer—if they are wearing the appropriate protective gear. Since the Sun's bright surface usually hides the corona, skywatchers will find this to be an exciting experience.

CHAPTER 1: THE SCIENCE BEHIND THE ECLIPSE

Partial, Annular, Total and Hybrid Eclipses

Partial Solar Eclipse: The partial eclipse occurs when the Sun, Earth, and Moon—the three friends in space—are not aligned in a straight line. Because of this unalignment, a partial solar eclipse is like the Moon taking a "bite" out of the Sun. Therefore, the Moon only partially covers the Sun rather than completely covering it. People not in the Moon's dark inner shadow can see this cool partial eclipse during a complete or annular eclipse.

Annular Solar Eclipse: When the Moon is not entirely covered and is at its apogee or most significant distance from Earth, it is known as an annular solar eclipse. Thus, you get this magnificent "ring of fire" encircling the Moon's surface. It is not the same as a total eclipse in which the Moon obscures the Sun. The term "ring of fire" eclipse refers to the appearance of the Moon during an annular eclipse, characterized by a bright ring surrounding the dark circle.

A Total Solar Eclipse: When the Sun is completely blocked by the Moon, a path of totality is created on Earth's surface. Those inside the path of totality will experience "daytime darkness" and be able to observe the Sun's corona. Individuals immediately outside of the path of totality will experience a partial solar eclipse.

Hybrid Solar Eclipse: Depending on their location, observers may see either a total or annular sun eclipse during a hybrid solar eclipse.

According to the educational website SpaceEdge Academy, 28% of solar eclipses are total, 35% are partial, 32% are annular, and only 5% are hybrid.

4. When do Eclipses Occur?

New Moon Phase: Solar eclipses can only occur during a new moon, the lunar phase, when the Moon is between the Earth and the Sun. This is because it's the only time the Moon can be directly in front of the Sun as viewed from Earth.

Eclipse Season: Due to the 5-degree tilt of the Moon's orbit relative to the Earth's orbital plane around the Sun, eclipses can only occur during certain times of the year, known as eclipse seasons. These seasons happen approximately every six months and last about 34 days. During each eclipse season, there is the potential for both a solar eclipse (when the Moon is between the Earth and the Sun) and a lunar eclipse (when the Earth is between the Sun and the Moon). The reason why we don't have eclipses every month is precisely due to this orbital inclination.

Saros Cycle: Eclipses occur in cycles, with the Saros cycle being the most famous. It's a period of approximately 18 years and 11 days, after which similar eclipses (with the Sun, Earth, and Moon in similar relative positions) recur.

Total solar eclipses are relatively rare events in any given location on Earth. On average, a total solar eclipse occurs at a specific location approximately once every 360 years. However, this can vary widely depending on the geographic location. Some regions may experience total solar eclipses more frequently than others.

5. Scientific Phenomena to Observe:

The visibility of the sun's Corona is the most dramatic and scientifically valuable aspect of a total solar eclipse. The corona is the outer atmosphere of the sun, which is usually obscured by the bright light of the Sun.

Just before and after totality, sunlight shining through the Moon's valleys creates bright spots known as Bailey's Beads surrounding the sun. This happens immediately before and after a total solar eclipse. It is fascinating that the origin of Bailey's Beads can be traced back to sunlight penetrating the craggy lunar surface. Certain regions of the Moon's surface, such as its craters, valleys, and mountains, block off sunlight, while others allow sunlight to shine through. Due to this illusion, the moon's edge appears to be made up of a necklace-like string of bright spots. Bailey's Beads are a brief and dynamic phenomenon that occurs for a few seconds when the Moon advances to obscure the Sun entirely or begins to move away from it. Named After Francis Baily: English astronomer Francis Baily first documented the beads in 1836 after seeing the occurrence during an eclipse. His story and description led to the moniker "Bailey's Beads." During a solar eclipse, Bailey's Beads appear and disappear to mark the start and finish of totality. For those who watch eclipses, they are among the highlights of the show. When only one bead is visible, it creates the stunning 'Diamond Ring' effect.

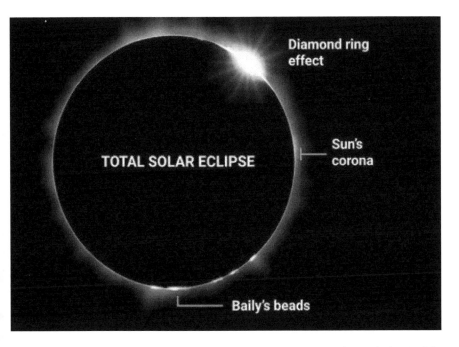

Solar prominences are flame-like projections from the Sun's surface, visible around the edges of the Moon during totality.

Changes in animal behavior include animals reacting to the sudden darkness of totality displaying unusual behavior that can be of interest to biologists.

Observers often note a noticeable drop in temperature during totality.

Mysterious, wavy lines of alternating light and dark, known as shadow bands, can sometimes be observed on the ground just before and after totality. These are caused by atmospheric turbulence.

6. The Impact of Eclipses on Solar Research:

Understanding the Solar Corona: Eclipses have been invaluable for studying the Sun's corona. The Sun's bright light usually obscures the corona, but during an eclipse, the sun's outer atmosphere provides crucial data about its structure and dynamics.

Solar Prominences and Magnetic Fields: Observations during eclipses have helped scientists study solar prominences and the Sun's magnetic fields.

Advancements in Astrophysics: Eclipse observations have contributed to our understanding of solar physics and the behavior of the Sun, which is crucial for understanding the broader universe.

Testing Scientific Theories: As mentioned earlier, solar eclipses have historically been used to test and validate scientific theories, such as Einstein's theory of General Relativity.

The 2024 solar eclipse offers a unique opportunity for both public enjoyment and scientific study, presenting phenomena that are fascinating to observe and valuable for understanding our Sun and its interactions with the Earth.

Now that we've uncovered the science let's delve into the history of solar eclipses and their impact on human culture.

Chapter 2: Historical Perspectives on Eclipses

The ancient Greek philosopher and mathematician Thales of Miletus is credited with accurately predicting a solar eclipse. This event had a profound impact on history.

In May 585 BC, two ancient kingdoms, the Lydians and Medes, had been at war for five years. These countries are now known as Turkey and Iran. Greek philosopher and mathematician Thales of Miletus, is credited with predicting the solar eclipse that occurred on that day, although his exact methods are unknown. According to historian Herodotus, the eclipse occurred during a battle between these two enemies. Both sides interpreted the sudden darkening of the sky as an ominous sign and a divine warning to stop the conflict. They negotiated a peace agreement, and the war was over.

This event is notable for several reasons: It demonstrates the early interest and understanding of celestial events in the ancient world; The story illustrates how astronomical events like eclipses were interpreted as divine or supernatural signs; and the prediction and its consequences had a direct impact on the course of history, influencing political and military decisions.

Thales's solar eclipse prediction is one of the earliest recorded instances of science intersecting with historical events, showing the profound impact astronomical phenomena can have on human affairs.

1. Eclipses in Ancient Times:

Throughout ancient history, eclipses have been observed and understood in various ways, and they have had a profound impact on astronomy, religion, and mythology.

Eclipse Records Throughout History:

China: A prevalent belief was that solar eclipses occurred when a celestial dragon or a dog devoured the Sun. This mythological interpretation led to various rituals and practices to chase away the creature and save the Sun. Lunar eclipses were similarly attributed to a dragon eating the Moon. In an effort to scare away the mythical creatures believed to be causing the eclipse, people would participate in loud, communal activities. These included banging pots and pans, setting off firecrackers, and playing loud music. The idea was to make enough noise to intimidate the creature into releasing the Sun or Moon.

The Chinese were meticulous in recording astronomical events. As early as the Zhou Dynasty (1046–256 BC), and certainly by the Han Dynasty (206 BC – 220 AD), official astronomers were tasked with predicting and documenting eclipses. The accuracy of these predictions was linked to the legitimacy and ability of the ruling dynasty, and failure to predict an eclipse could have serious ramifications. The Emperor, seen as the Son of Heaven and a mediator between Heaven and Earth, was expected to perform certain rituals to prevent or alleviate the ill effects of an eclipse. This often involved fasting, performing sacrifices, and public acts of righteousness. It is said that two court astronomers, Hsi and Ho, were put to death for their inability to foresee an eclipse in or about 2134 BC.

These celestial events spurred advancements in Chinese astronomy. The need to predict eclipses accurately led to improvements in astronomical knowledge and techniques. For example, the famous Chinese astronomer and mathematician Shìxiàn made significant contributions to understanding lunar and solar eclipses.

CHAPTER 2: HISTORICAL PERSPECTIVES ON ECLIPSES

Babylon: Eclipses for the Babylonians were often seen as omens, particularly pertaining to the fate of the king and the country. They were interpreted as divine messages, and their occurrence required careful attention and interpretation. Babylonian scholars, particularly the priests known as Chaldeans, developed an elaborate system of omen-based astrology. They believed that the gods communicated through natural events, and eclipses were among the most important. Detailed records of eclipses were kept, which were used to predict future events and advise the king. The Babylonians were meticulous in their astronomical observations and maintained detailed records of celestial events, including eclipses. Tablets from the period, such as the famous Enuma Anu Enlil, contain eclipse records and omen texts. Their observations were so precise that they could predict lunar eclipses with a fair degree of accuracy.

One of the most significant contributions of Babylonian astronomy was the discovery of the Saros cycle, an 18-year period after which eclipses repeated. This understanding allowed them to predict solar and lunar eclipses with considerable accuracy. When an eclipse was predicted, especially if it was deemed to be an ominous sign, the Babylonians engaged in various rituals to avert disaster. These might include increased offerings to the gods, prayers, and other religious ceremonies. Often, a substitute king was installed temporarily to bear any negative consequences predicted by the eclipse, protecting the real king from harm. The Babylonians developed a form of mathematical astronomy, which was quite advanced for its time. They used arithmetic to predict celestial events, including eclipses, and their methods laid the groundwork for later astronomical developments in the Hellenistic world and beyond.

Greece: Ancient Greek philosophers, such as Anaxagoras and Thales, were among the first to propose naturalistic explanations for eclipses. They understood that solar eclipses occurred when the Moon passed between the Earth and the Sun, causing a temporary darkening of the Sun's light. On the other hand, Lunar eclipses happen when the Earth comes between the Sun and the Moon, casting a shadow on the Moon.

Eclipses helped the ancient Greeks comprehend celestial mechanics. For instance, Aristarchus of Samos calculated the relative sizes and distances between the Sun and Moon using a lunar eclipse.

While there was some scientific understanding of eclipses, they were also viewed with fear and superstition. The sudden darkening of the sky during a solar eclipse was often associated with the wrath of the gods, and people would sometimes engage in rituals to ward off the perceived danger. Eclipses were often recorded in ancient Greek history. For example, the eclipse of 585 BCE, known as the Eclipse of Thales, was accurately predicted by the philosopher Thales. This event marked an early success in predicting celestial phenomena and contributed to the development of astronomy.

Eclipses could also carry symbolic meanings in ancient Greek culture. They were sometimes interpreted as omens or signs from the gods, foretelling important events or societal changes. Some Greek astronomers and scholars actively observed and studied eclipses despite the superstitions. They made efforts to predict their occurrences and understand the celestial mechanics behind them.

Eclipses in Mythology and Religion:

Greek mythology often incorporated eclipses into stories. One famous myth is the story of the Titan Rhea, who saved her son Zeus by hiding him during a celestial battle, resulting in an eclipse. This myth connected eclipses with divine intervention.

According to Norse mythology, eclipses occur when the Sun and Moon are caught and consumed by the sky wolves, Skoll and Hati. People would make loud noises to frighten these wolves away. In the Norse myth of Ragnarök, a series of apocalyptic events that mark the end of the world, there is a reference to a great wolf named Sköll, who is said to chase the Sun. It is prophesied that during Ragnarök, Sköll will finally catch and devour the Sun, plunging the world into darkness. While this is not a direct reference to a solar eclipse, it resembles the imagery of the Sun being temporarily obscured.

In Hindu mythology, eclipses are often attributed to the activities of two mythological demons, Rahu and Ketu. According to the legend, these demons disguised themselves as gods to taste the nectar of immortality. When their deception was discovered, Lord Vishnu decapitated Rahu, but because he had consumed the nectar, his head remained immortal, becoming Rahu, and his body became Ketu. They are said to chase the Sun and the Moon to exact their revenge, causing solar and lunar eclipses when they catch them. This legend explains the Hindu belief in the importance of taking precautions during eclipses and performing rituals to protect oneself from Rahu and Ketu's evil influence.

In Hindu culture, eclipses are considered inauspicious times, and certain rituals are followed to maintain purity and ward off negative energies. Many people observe fasting and meditation during eclipses. After the eclipse, it's common to take a ritual bath in sacred rivers or visit temples to cleanse oneself spiritually.

In Navajo mythology, there is a story about a bear responsible for causing a solar eclipse. According to the legend, the bear gets angry with the Sun and takes a bite out of it, causing it to disappear temporarily. The Navajo people perform rituals during solar eclipses to ensure the safe return of the Sun. These rituals involve singing, praying, and making offerings to appease the bear and encourage the Sun's return.

In various Pueblo and Hopi tribes, eclipses are seen as important events with spiritual significance. Some tribes believe that eclipses represent a transformation, cleansing, and renewal time. Rituals and ceremonies are often conducted during eclipses to purify and protect the community. For example, the Cahuilla and Crow tribes have traditions related to eclipses, including singing and dancing, to ward off negative influences.

Eclipses are viewed as opportunities for prayer and reflection among the Oglala Sioux. During a solar eclipse, they believe that the Sun is temporarily weakened, and this is seen as a time for individuals to seek spiritual guidance and strength through prayer and meditation.

It's important to note that Native American myths and legends are diverse, and the beliefs surrounding eclipses can vary widely among different tribes and communities. These legends reflect the deep connection that Native American cultures have with nature and the cosmos and often emphasize the importance of rituals and ceremonies to maintain harmony and balance in the world.

These examples show how eclipses were significant astronomical events in ancient times and integral to various cultures' mythology, religion, and scientific progress. Their impact on early astronomy was particularly profound, aiding in the development of the earliest models of the cosmos and celestial mechanics.

2. Modern Historical Events:

Einstein's Theory of General Relativity (1919): The acceptance of Einstein's theory was significantly influenced by the total solar eclipse that occurred on May 29, 1919. Astronomer Arthur Eddington saw the bending of starlight around the Sun during the eclipse, which General Relativity predicted. The results, which defied Newtonian physics predictions, were widely reported and were viewed as a significant validation of Einstein's theory.

The discovery of helium (1868): French astronomer Jules Janssen and English astronomer Joseph Norman Lockyer made history on August 18, 1868, when they spotted a yellow spectral line in the Sun's chromosphere during a solar eclipse. They postulated the presence of a newly discovered element, subsequently dubbed helium, which wasn't found on Earth until 1895.

On August 21, 1914, a total solar eclipse took place, coinciding with the outbreak of World War I. The Middle East and Eastern Europe were on the totality's path. According to some modern accounts, there was a greater sense of dread and uncertainty at the beginning of the war due to the eclipse's darkened skies.

On June 24, 1778, a total solar eclipse was visible over most of North America, including major Revolutionary War battlegrounds. Although it did not directly affect how the battle turned out, it was mentioned in both soldiers' and civilians' diaries.

The British astronomer and broadcaster Sir Patrick Moore (1923–2012) was well-known for his passion for eclipses. He saw around 70 solar and lunar eclipses while traveling the world to see them. Through his writings and television shows, Moore frequently discussed his experiences, igniting the interest of many in astronomy.

CHAPTER 2: HISTORICAL PERSPECTIVES ON ECLIPSES

The "Eclipse Chaser," David Makepeace, has seen solar eclipses while visiting every continent—including Antarctica. His stories frequently center on the spiritual and transforming aspects of seeing a total solar eclipse, characterizing the event as compelling and transformative.

2017 Total Solar Eclipse in the United States: Due to its passage across the whole continental United States, the 2017 solar eclipse was one of the most watched and photographed eclipses in history. According to a NASA assessment, an estimated 1.85 to 7.4 million people flocked to the path of totality to witness the eclipse. Visitors to states in the path of totality, such as Oregon, Idaho, Wyoming, and Missouri, increased significantly. Reservations for hotels, campgrounds, and other lodging were frequently made months or even years in advance. This had a significant economic impact. For instance, the state of Oregon saw an influx of about a million tourists, who spent millions on lodging, food, travel, and eclipse-related goods, boosting the local economy.

Another noteworthy event that drew a lot of tourists was the 1999 Total Solar Eclipse in Europe. It went through many nations, including Romania, Germany, France, and the United Kingdom. Millions of people are thought to have traveled throughout Europe to witness the path of totality, with significant numbers of people arriving in parts of the UK like Cornwall.

These illustrations demonstrate the profound effect solar eclipses have on historical occurrences and scientific knowledge. The anecdotes of eclipse chasers also shed insight into the individual and frequently significant experiences of those who follow these astronomical phenomena around the world.

3. Personal Stories from the Past:

Monica Young described her experience with a solar eclipse, noting the strange, silvery sunlight, the nervous behavior of dogs, and the sudden onset of darkness. As the eclipse reached totality, she was struck by the awe-inspiring sight of the black hole in the sky, surrounded by the ethereal glow of the Sun's corona. She vividly recalls the diamond ring effect, the bright prominence of Venus, and the rapidly passing moment of totality.

Clinical psychologist Kate Russo detailed the primal fear induced by the creeping darkness of a solar eclipse. As totality occurs and darkness descends, a sense of awe overwhelms onlookers. This awe stems from the vastness and power of the universe, leading to feelings of insignificance and a profound sense of connection with humanity and nature. The experience concludes with a euphoric desire to repeat the experience, showcasing the intense emotional journey of witnessing a total solar eclipse.

Annie Jump Cannon: The famous astronomer, known for her work in classifying stellar spectra, was deeply moved by her experience of viewing a solar eclipse in 1896. This event is said to have cemented her passion for astronomy.

A 2017 U.S. solar eclipse viewer described the event as "a moment of celestial majesty and beauty." They mentioned feeling a deep sense of connection to the universe and a renewed appreciation for the wonders of nature.

In Casper, Wyoming, an eclipse experience was described as being part of an electric atmosphere with eclipse chasers worldwide. The small town transformed into a hub of excitement and festivity. Despite some anxiety about the cloud cover, the event turned out to be a spectacular success as totality approached. This narrative highlights the communal aspect of the eclipse, with people sharing the awe and joy of the experience. Having planned this trip for two years, the writer reflected on the incredible payoff of witnessing such a remarkable event and looked forward to future eclipses.

A Catalyst for Change: Witnessing an eclipse becomes a transformative experience for some. For example, a person who viewed the 1999 total solar eclipse in Europe described it as a life-altering event that led them to reevaluate their life priorities and careers, ultimately leading to a more fulfilling path.

Inspiration for Artists and Writers: Eclipses have inspired numerous works of art and literature. Many artists and writers have cited eclipses as key influences in their work, using the imagery and emotion evoked by eclipses to fuel their creativity.

Present Day: Solar eclipses are well understood today, and their occurrences can be accurately predicted. They continue to be a subject of interest for both professional astronomers and the public for their beauty and the unique opportunities they provide for solar observation.

Eddington and Relativity: As mentioned earlier, the 1919 eclipse expedition led by Arthur Eddington confirmed Einstein's theory of General Relativity and bolstered Eddington's career. This pivotal moment in astrophysics inspired generations of scientists to pursue careers in the field.

Young Astronomers: Modern solar eclipses continue to inspire young people to pursue science. For instance, following the 2017 eclipse in the U.S., many STEM education programs reported increased interest in astronomy and physics from students who had witnessed the event.

These examples illustrate the profound and varied impact that seeing a solar eclipse can have. From personal transformation and artistic inspiration to significant contributions in the field of science, eclipses have left an indelible mark on many lives.

Thus, the cultural significance of solar eclipses is vast and varied, embodying a mix of fear, reverence, curiosity, and celebration. These events have shaped religious and cultural practices and have been instrumental in advancing our understanding of astronomy and timekeeping.

"With a grasp of eclipse history, let's transition to preparing for the 2024 event, starting with choosing the best location."

Chapter 3: Finding Your Perfect Spot

An estimated one to four million people are expected to travel to see the total solar eclipse on April 8, 2024, in the United States.

1. Choosing Your Location

Choosing an optimal location for viewing the 2024 solar eclipse involves several factors, and you can also gain insight from recommended spots and narratives from locals in prime locations. Here's a breakdown to help guide your decision:

Factors to Consider:

- Crowd Potential: Popular viewing spots can become very crowded. Decide if you prefer a more solitary experience or are okay with being with large groups of people. Larger crowds can offer a more communal experience but might also mean more traffic and fewer accommodations available. Expect traffic congestion on the way to and from your eclipse viewing location, especially if it's a popular spot. Plan your travel routes and departure times accordingly.
- Check Weather Conditions: Monitor the weather forecast for your chosen

eclipse viewing location. Clear skies are crucial for a successful eclipse viewing, so consider having a backup plan in case of bad weather. Select a viewing location with minimal light pollution and a good chance of clear weather. Some popular eclipse destinations include national parks, remote areas, and observatories.

- Duration of Totality: Some locations along the path of totality will experience a longer duration of total eclipse. You may want to choose a location that maximizes this duration time.
- Arrive Early: Arrive at your selected location well before the eclipse's start time. This lets you set up your equipment, familiarize yourself with the surroundings, and avoid any last-minute rush.
- Enjoy the Moment: During the eclipse, take time to enjoy the experience. Eclipse events are not only visually stunning but also emotionally moving. Capture photographs and memories, but remember to immerse yourself in the moment.
- Respect the Environment: Be mindful of the environment and local regulations. Clean up after yourself, respect local customs and traditions, and leave no trace of your visit.
- Travel Distance: Choosing your viewing location will depend on how far you are willing and able to travel. Traveling by car, plane, or train is a consideration.
- Total or Partial Eclipse: You'll need to decide if you want the full total eclipse viewing or if you are okay with seeing the partial eclipse. Where you go to see the eclipse will determine how much of the eclipse you are able to see.
- City or Country: Your experiences will look different depending on whether you are viewing the eclipse in the city or the country. Imagine being on a high-rise roof-top in a major city or on a country road in Texas. Either will yield a spectacular experience.
- Share the Experience: Eclipse viewing can be a communal experience. Share your knowledge and equipment with fellow eclipse enthusiasts, and consider joining eclipse-related events or groups to enhance your experience

Recommended Viewing Spots:

- Texas: Places like San Antonio and Dallas are expected to be good viewing spots. Texas generally has favorable weather conditions and offers both urban and rural viewing options.
- Midwest: Carbondale, Illinois, is notable as it is near the point of greatest duration for the 2024 eclipse. It also experienced the 2017 eclipse, making it a unique spot for eclipse enthusiasts.
- Eastern U.S.: In the eastern United States, locations in Ohio, New York, and Vermont could offer good viewing experiences, with the added charm of their natural landscapes.

The map showing the Path of Totality is the best place to find the general scope of the total solar eclipse event. Don't worry about the center line on the map as the total eclipse can be seen in the entire total eclipse band of the map.

Resources for more information

- Community Forums and Blogs: Check out local community forums, Chamber of Commerce, blogs, or social media groups where residents of towns along the path of totality might share their plans and expectations for the eclipse. This can give you a sense of the community's enthusiasm and any local events being planned.
- Local News Coverage: Local newspapers or TV stations may feature stories on how their towns are preparing for the eclipse, offering insights into the local vibe and any special viewing parties or festivals.
- Personal Testimonials: Websites dedicated to eclipse chasing or astronomy might have testimonials from people who have experienced previous eclipses. These narratives can provide a unique perspective on what it's like to be in a prime location during the event.

Remember, the best location for you will depend on your personal preferences regarding these factors. Planning ahead is crucial, especially for accommodation and transportation, as the best spots can get booked up quickly. Keep an eye on local resources and enthusiast communities for the most current information and tips.

2. Choosing Your Location

The best viewing conditions within the totality path depend on the weather. Clear skies are ideal for observing the total eclipse. Local weather patterns and historical data can help predict viewing conditions but can never be guaranteed.

Weather: Check historical weather data for potential locations to gauge the likelihood of clear skies. Regions with fewer average cloud cover and lower precipitation in April are preferable. Websites like the National Weather Service or local meteorological services can provide this information.

Accessibility: Consider how easy it is to travel to and within the viewing location. Some areas may be very remote with limited access, while others might be near major cities or highways. Also, consider the availability of accommodations and facilities.

3. Travel and Accommodations

Due to the narrowness of the path, planning is essential for those wanting to view a total solar eclipse. This includes considering transportation, accommodation, and potential crowds, especially in areas with limited infrastructure.

CHAPTER 3: FINDING YOUR PERFECT SPOT

Planning Your Visit (Accommodation, Activities):

- Book Early: Accommodations in the path of totality can fill up quickly, sometimes years in advance. Consider hotels, Airbnb, camping sites, RV parks, State or National Parks, or even local rentals. Use sites like Expedia or Hotels.com to find hotels in your targeted location. Hotel Chains may be a good source for finding a hotel with vacancies in an out-of-the-way location, e.g., Hyatt, Marriott, etc.
- Planning a Gulf of Mexico Cruise or a plane flight on the eclipse day have been suggested as an innovative way to celebrate the 2024 Eclipse.
- Eclipse Events: Check for any local eclipse events or gatherings planned. These could include educational talks, events at a zoo, viewing parties, nature hikes, museums, or community festivals that would enrich your visit.
- Consider making this a family event with a family reunion or family meeting for the day with activities for all ages.

Getting Involved in the Community:

- Local Companies: Give local companies your support when you're there. Go to local stores, eat at local eateries, and interact with local service providers.
- Cultural Events: If any cultural events are available, participate in them. This may be a really beneficial approach to comprehend and value the neighborhood.
- Community Involvement: Talk to the locals to find out about their views on the eclipse as well as any customs or stories they may have. Making genuine connections in this way can be quite beneficial.

Environmental and Safety Considerations:

- Eclipse Eyewear: To view the eclipse safely, make sure you have eclipse glasses with an ISO certification. It is not safe to watch solar eclipses with only regular sunglasses. Follow safety guidelines when observing the eclipse. Do not try to view the eclipse through a camera or binoculars without proper filters.
- Pack Essentials: Prepare for your trip by packing essentials like water, food, sunscreen, comfortable clothing, and any necessary camping gear if you plan to stay overnight. Be self-sufficient, especially if you'll be in a remote area.
- Traffic Safety: On the day of the eclipse, be ready for heavy traffic. To avoid the crowd, aim to reach your viewing spot early and remain there for some time following the eclipse.
- Environmental Impact: Consider the effects you have on the environment. Respect wildlife and natural environments, stay on authorized trails and observation places and dispose of rubbish appropriately.
- Weather Preparedness: Be ready for a range of meteorological circumstances. Pack water, hats for sunny weather, sunscreen, and rain gear just in case.
- Health Precautions: Be prepared for any health issues like allergies or insect bites depending on the area and season. Keeping a first aid kit on hand is also a good idea.

Recall that the ideal place for you will rely on your individual tastes in relation to these elements. It is essential to plan ahead, particularly for lodging and transportation, since the best locations tend to fill up quickly. For the most up-to-date information and advice, keep an eye on regional resources and enthusiast forums.

Here are some recommendations for comprehensive sources of information on the 2024 solar eclipse, including maps, accounts of the viewing experience, and histories of the towns along the path:

CHAPTER 3: FINDING YOUR PERFECT SPOT

Maps of the 2024 Path in Detail:

- The official website of NASA: Comprehensive solar eclipse information is available from NASA, including interactive maps that display the eclipse's path. Their maps often provide information on timings, path width, and totality duration at different sites.
- Website of the Great American Eclipse: With a focus on American eclipses, this website provides comprehensive state-by-state breakdowns, maps, and instructions for the 2024 eclipse.
- Time and Date: Local times for the eclipse's beginning and ending in various regions are provided by this website, along with interactive maps and comprehensive information about the eclipse's path.

How the Path Impacts the Experience of Watching:

- Websites and Magazines about Astronomy: Articles in magazines like "Sky & Telescope" or "Astronomy" frequently discuss how the route of totality impacts the viewing experience, taking into account meteorological conditions and topographical features.
- Local Astronomical Groups: Many local astronomical groups can tell you how viewing is affected by the eclipse path in different places. They can provide information on what to expect and the best local viewing locations.
- Science museums and planetariums: These establishments frequently host lectures and other events in the run-up to an eclipse, during which they talk about how the path affects the viewing experience.
- Stories from Towns Along the Path: Local News Sources: Towns along the 2024 path will probably have articles and features about their communities' eclipse preparations published in local newspapers and news websites.
- Websites and Blogs on Travel: Seek out travel-related blogs or websites that may have articles or itineraries about communities along the eclipse route, emphasizing the special experiences these places have to offer.
- Social media: Websites such as Facebook, Instagram, and Twitter can be

excellent sources of up-to-date information about plans and stories from the towns along the route. You can find updates and anecdotes from local community groups or pages about the eclipse.

As the 2024 eclipse draws near, it will be helpful to check these sources frequently for the most up-to-date and comprehensive information. Whether it's the local human-interest stories from news sites, the technical and astronomical facts from NASA, or the helpful viewing advice from astronomical groups, each source gives a different viewpoint.

Recall that careful preparation can turn your eclipse viewing experience into a fun and educational adventure and a breathtaking moment of celestial wonder. You will have a better experience and make a positive influence if you interact with local communities and environs courteously and responsibly.

With your location set, let's equip you with the knowledge for safe and effective eclipse viewing.

Chapter 4: The Art (and Safety) of Eclipse Viewing

"A teenage amateur astronomer named Sophie was excited to see this celestial event during the amazing solar eclipse of 1999 that swept over Europe. Equipped with improvised telescopes, she looked up, waiting for the total eclipse. However, since she was too excited and didn't have the required safety equipment, Sophie damaged her retina, which left her right eye permanently blind. This tragedy is a sobering reminder of how crucial eclipse safety is. It serves as a reminder that, although breathtaking, solar eclipses should also be treated with respect and prudence. The Sun is mighty, and its rays can still be dangerous during an eclipse. We must use safe eclipse glasses or indirect viewing techniques to preserve our priceless gift of sight. In addition to being a tale of astronomical passion, Sophie's story serves as a sobering reminder of the thin line that separates awe and risk in the face of the grandeur of nature."

This anecdote highlights the importance of viewing solar eclipses carefully, even though they are a beautiful natural phenomenon. To thoroughly enjoy the experience without endangering our health, it is imperative to prioritize safety as the Sun's rays, even when partially hidden, can cause severe eye damage.

1. Safety Procedures

When seeing a solar eclipse, safety is crucial. During any part of an eclipse, avoid looking directly at the Sun without using appropriate eye protection, such as portable solar viewers or certified solar eclipse glasses. Ordinary sunglasses, do-it-yourself filters, or other remedies are unsafe and can seriously harm your eyes. Making a pinhole projector, which projects an image of the eclipse onto a surface without requiring one to stare at the Sun, is another safe method to see the eclipse. Furthermore, ensure that any binoculars or telescopes you plan to use to see the eclipse are equipped with solar viewing equipment or have specific solar filters. By following these precautions, you may take in the breathtaking grandeur of a solar eclipse without having to worry about damaging your eyes.

2. Safety Gear

Kinds of Equipment for Safe Viewing:

- Eclipse glasses are specialized eyewear with solar filters that block harmful light. They must meet the international safety standard ISO 12312-2. Remember that a genuinely safe solar viewer does more than reduce the Sun's visible light to a comfortable brightness level. It also blocks potentially harmful UV and IR radiation. Conventional sunglasses are unsafe for seeing the Sun. Note that special-purpose solar filters (eclipse glasses) are at least 1,000 times darker than ordinary sunglasses!
- Pinhole projectors: This easy do-it-yourself technique projects the sun's image onto a flat surface through a tiny hole. It lets you see the eclipse through an indirect line.
- Solar Viewers: A unique substance that filters sunlight is built into these

portable gadgets. They have to adhere to international safety regulations, much like eclipse glasses.
- Telescopes with Solar Filters: Telescopes with solar filters can be used by anyone who wants a closer look. To safeguard the apparatus and your eyes, these filters must be positioned on the front of the telescope.

The Scientific Basis for Safety's Importance

- Retinal Damage: The retina, the light-sensitive layer at the back of the eye, can be burned by the sun's visible solid and UV rays. Since the retina lacks pain receptors, damage can occur without any sensation of pain.
- Cumulative Effect: Damage can result from even a brief exposure to the sun without appropriate protection. The more you do it, the more likely it is that the damage will accumulate.
- Afterimage: Solar retinopathy, which is caused by unprotected sun viewing, can cause visual abnormalities such as afterimages, blurriness, or even a momentary or permanent blind spot in the middle of the vision.

3. Individual Testimonials of Dangerous Viewing Outcomes:

Retinal Damage Case: A witness from a previous eclipse described what it was like to look at the Sun briefly without any safety gear. Their ability to read and recognize faces was significantly impacted, and they were left with a permanent blind patch in their central vision.
-

Cameraman's Remorse: An account of taking a picture of an eclipse using a camera without a solar filter was once told by a photographer. The intense light harmed his eyes and the camera's sensor, resulting in long-term vision problems.

Childhood Experience: There is a legend of a person who, fascinated by the spectacle as a child, peered out of protection at a solar eclipse. This resulted in a lesson about the significance of eclipse safety as well as a lifetime battle with distorted eyesight.

These arguments emphasize how crucial it is to see solar eclipses safely by utilizing the appropriate tools and techniques. Prioritizing eye safety during these uncommon celestial phenomena is crucial since retinal damage from the Sun's beams is permanent.

4. Using tools and apps for accurate viewing

Eclipse Apps: Many smartphone apps have been developed specifically for observing eclipses, offering GPS location services, exact time, and eclipse phase reminders.

- Solar Eclipse Timer: This app provides precise timing for all phases of a solar eclipse, including contact times for partial and total phases. It offers audio and visual cues to help you maximize your eclipse viewing experience.
- Eclipse Safari: Eclipse Safari is a comprehensive app that offers eclipse-related information, including eclipse maps, weather forecasts, and tips for safe viewing. It's a valuable resource for planning your eclipse trip.
- Star Walk 2: Star Walk 2 is an astronomy app that can help you identify stars, planets, and constellations during an eclipse event. It's a handy tool for stargazing before and after the eclipse.
- SkySafari: SkySafari is a powerful astronomy app that provides detailed information about celestial objects, including the Sun and Moon. You can use it to track the positions of the Sun and Moon during an eclipse and identify other celestial objects in the sky.
- Solar Eclipse by Redshift: This app offers eclipse simulations, countdown

CHAPTER 4: THE ART (AND SAFETY) OF ECLIPSE VIEWING

timers, and interactive maps to help you plan your eclipse viewing. It provides real-time updates and information about upcoming eclipse events.

- Lunar Eclipse Guide: For lunar eclipses, the Lunar Eclipse Guide app provides information about upcoming lunar eclipse events, visibility maps, and details about the phases of the eclipse.
- Clear Outside: Weather conditions are crucial for eclipse viewing. Clear Outside is a weather forecasting app that provides detailed cloud cover and visibility predictions for your location, helping you choose the best spot for viewing.
- Eclipse Soundscapes: This app is designed for visually impaired eclipse enthusiasts. It provides audio descriptions and real-time sonification of the eclipse, allowing everyone to experience this celestial event.

Remember to download and familiarize yourself with these apps well in advance of the eclipse event to make the most of your viewing experience. Additionally, always prioritize safety by using proper eye protection when observing a solar eclipse.

Apps for photography

Some apps assist users with camera settings and timing so they can take good pictures of the eclipse.

- Eclipse Megamovie: If you're interested in contributing to citizen science during an eclipse, the Eclipse Megamovie app allows you to record and submit photos and videos of the eclipse to create a crowd-sourced movie of the event.
- PhotoPills: PhotoPills is a versatile photography planning app that can help you calculate the best times and camera settings for capturing eclipse photos. It provides information on eclipse timing, including contact times and the duration of different eclipse phases.

- Solar Eclipse Timer: While primarily focused on timing eclipse phases for viewing, Solar Eclipse Timer can also be beneficial for photographers. It provides audio and visual cues to help you capture the most critical moments during a solar eclipse, such as totality.
- EclipseDroid: This app is designed specifically for eclipse photography. It calculates exposure settings, offers a solar filter calculator, and provides a countdown timer for eclipse events. EclipseDroid is a helpful tool for planning your eclipse photography.
- Eclipse Calculator and Sundroid Pro: These apps offer eclipse timing information, including contact times and the position of the Sun and Moon during an eclipse. They also provide data on the altitude and azimuth of celestial objects for better composition planning.
- Golden Hour: Knowing the exact timing of the golden hour (the period shortly before sunset or after sunrise) is crucial for eclipse photography. Golden Hour apps help you determine when the lighting conditions will be most favorable for your shots.
- Photographer's Ephemeris: This app provides information about the position of the Sun and Moon at any given time and location. It's useful for planning the alignment of the eclipse with landmarks or natural features.

These photography apps offer a range of tools and features to assist eclipse photographers in capturing memorable and well-exposed eclipse photos. Remember to practice using these apps and your camera settings before the eclipse event to ensure a successful photography session. Additionally, always use proper solar filters or eclipse glasses and camera equipment to protect your eyes during solar eclipse photography.

CHAPTER 4: THE ART (AND SAFETY) OF ECLIPSE VIEWING

5. Eclipse timing and phases.

To make the most out of your solar eclipse viewing experience, it can be very beneficial to know the eclipse's schedule and phases, use precision equipment and applications, and draw lessons from previous memorable experiences:

When and how the eclipse occurs:

- Recognizing Phases: There are three distinct phases to a solar eclipse: the partial phase, which occurs when the Moon partially covers the Sun, the total phase, which occurs when the Sun is fully covered, and the reversal of these phases. Every stage presents a unique visual encounter.
- Verifying Local Times: Location-specific differences exist in the eclipse phases' exact timing. Use internet or local astronomical resources to get the exact times for your location. This will assist you in organizing how you will see the ending, the totality, and the partial phases.
- Totality Duration: The complete phase's duration is especially significant. There will be a lengthier totality time in some places. It's possible to significantly improve the experience by placing yourself in these regions.

6. Interesting Eclipse Activities

Construct pinhole projectors for a straightforward and safe indirect view of the eclipse.
  Step-by-step instructions:
  Materials You'll Need:

- A cardboard or heavy paper box (e.g., a cereal box or shoebox)
- Aluminum foil
- A small piece of white paper or poster board

- Tape
- A pin or needle
- A pair of scissors or a utility knife
- A ruler
- A pencil
- Black construction paper or paint (optional)

Step-by-Step Instructions:

- 1. Prepare the Box: Take your cardboard or paper box and remove any flaps or openings, leaving just one end open. If the box has a shiny or reflective interior, you may want to paint it black to reduce glare and improve contrast.
- 2. Create a Pinhole: On one of the shorter sides of the box (the bottom when the box is standing upright), use a pencil to make a small dot in the center. This will be your pinhole.
- 3. Cover One End with Aluminum Foil: Carefully tape a piece of aluminum foil over the open end of the box. Ensure it's smooth and flat. This will be your projection screen.
- 4. Cut a Viewing Hole: On the opposite end of the box (the top when the box is standing upright), cut a rectangular viewing hole. This hole should be at least a few inches wide and several inches long. If you'd like, you can also cover this hole with a piece of black construction paper to create a dark interior for better visibility.
- 5. Create a Projection Surface: Inside the box, at the end with the viewing hole, tape a small piece of white paper or poster board to the side opposite the pinhole. This will be where you see the projected image.
- 6. Position the Box: Stand with your back to the Sun and hold the box with the pinhole facing the Sun. Adjust the angle until you see a small, focused image of the Sun on the white paper inside the box. Be patient and make small adjustments until the image is clear.
- 7. Observe the Eclipse: As the eclipse progresses, you will see the image of the Sun change shape, with the Moon slowly covering or uncovering

it. Enjoy the eclipse safely by looking at the projection on the white paper inside the box. Do not look directly at the Sun without proper eye protection.

A pinhole eclipse viewer is a simple and effective way to watch a solar eclipse without risking eye damage. Just make sure to set up your viewer a few minutes before the eclipse begins so that you're ready to observe the event safely.

Shadow Experiments:

You can conduct a simple and fascinating shadow experiment during a solar eclipse to observe the effects of the eclipse on shadows. This experiment will help you see how the changing position of the Sun during an eclipse affects the shapes and sharpness of shadows. Here's how to do it:

Materials You'll Need:

- A flat, open area with a clear view of the Sun and a surface where shadows can be cast (e.g., a sidewalk, a white sheet of paper on the ground, or a wall)
- A tall object like a tree or a lamppost

Step-by-Step Instructions:

- 1. Choose a Location: Find a suitable location where you can observe the ground or a vertical surface where shadows will be cast. Ensure you have a clear and unobstructed view of the Sun during the eclipse.
- 2. Set Up the Experiment: Before the eclipse begins, select a tall object (e.g., a tree or lamppost) that will cast a shadow on your chosen ground or surface. Position yourself so that you can observe this object and its shadow.
- 3. Observe the Initial Shadow: Before the eclipse starts, note the tall object's shadow. Pay attention to its shape, length, and sharpness. This

will serve as your baseline shadow.
- 4. Monitor the Shadow Changes: As the solar eclipse progresses, regularly check the shadow cast by the tall object. You'll notice several exciting changes:
- The shadow will become sharper and more defined as the eclipse progresses toward totality (if applicable to your location). This is because the Moon partially blocks the Sun's light source, creating a smaller and more focused light point.
- The shape of the shadow may change. During a total solar eclipse, you might see the shadow take on a crescent shape, mirroring the shape of the eclipsed Sun.
- Pay attention to any unusual or interesting effects in the shadow, such as variations in sharpness and clarity.
- 5. Take Notes and Photos: Document your observations by taking notes and photographs throughout the eclipse. This will help you record the changes in the shadow's appearance and how they correlate with the eclipse phases.
- 6. Share Your Findings: After the eclipse, you can share your observations and photos with friends, family, or eclipse enthusiast communities. It's a great way to engage others and discuss the fascinating effects of eclipses on shadows.

This shadow experiment provides a unique opportunity to witness firsthand the impact of a solar eclipse on the quality and characteristics of shadows. It's a fun and educational way to explore the changes in natural phenomena during these celestial events.

Sun and Moon Art: Draw inspiration for your artwork from the eclipse. This might entail making digital art, sketching, painting, or coloring using an eclipse as inspiration.

Taking Pictures of the Eclipse: Taking pictures of the eclipse can be a gratifying and demanding endeavor for photographers. Using specialized sun filters to

CHAPTER 4: THE ART (AND SAFETY) OF ECLIPSE VIEWING

take pictures of eclipses, learn what camera settings work best.

Throwing a viewing party or attending one:

- Community Viewing Events: Find out whether there will be a viewing event held by the astronomy club, museum, or school in your area. Experts are frequently present at these sessions, making it an excellent opportunity to learn while viewing the eclipse.
- Plan a Neighborhood watching Party: If you own a decent location for watching, consider throwing a party. Offer instructional resources, eclipse glasses, and possibly even refreshments with a theme.
- Themed Parties: Organize a viewing party with décor, activities, and cuisine based on solar and lunar themes. This can be a memorable way for children to make the occasion.
- Online Viewing Parties: If you reside somewhere that is not in the path of totality, you might want to think about hosting or attending an online viewing party where you can watch live feeds of the eclipse.

Possibilities for Children and Families to Learn:

- Eclipse Workshops: In the days preceding an eclipse, many science museums, observatories, and educational institutions host workshops. These can involve things like constructing sun viewers, studying eclipse science, and being aware of safe viewing procedures.
- Astronomy Camps: See if nearby astronomy camps or programs offer eclipse-focused special sessions. For kids, these can be entertaining as well as educational.
- Encourage kids to write a drama, song, or poetry about the eclipse to perform for loved ones.
- Storytelling Sessions: Host or go to storytelling events with a mythological, legendary, or scientific bent about eclipses. This might be an enjoyable method to discover how eclipses are seen in various cultural contexts.
- Simple science activities that illustrate the fundamentals of an eclipse can

be exciting and instructive—for instance, modeling the Earth, Moon, and Sun with balls to demonstrate how eclipses happen.

These events heighten the anticipation for the eclipse and offer a forum for education and community involvement. They have the power to elevate the eclipse-watching experience for people of all ages to a more social and informative level. Making the most of the eclipse experience requires careful planning based on location and timing, utilizing the appropriate equipment for photography and assistance, and getting ideas from other people's experiences. With enough planning, a routine viewing can become an unforgettable, once-in-a-lifetime experience.

Having covered how to view the eclipse safely, let's turn our lenses skyward to capture its majesty through photography.

Chapter 5: Eclipse Photography for Beginners

"The night sky is a canvas adorned with stars, and each eclipse is a stroke of cosmic artistry." - Unknown

1. Essential Photography Tips

Choosing the right equipment.

- Camera: Use a digital camera with manual settings. DSLR (Digital Single-Lens Reflex) cameras or mirrorless cameras are ideal for eclipse photography due to their manual control options.
- Lens: A telephoto lens with a focal length of at least 200mm is recommended for capturing details of the Sun during the eclipse. A longer lens, such as 400mm or 600mm, will allow for even closer shots.
- Tripod: Use a sturdy tripod to ensure stability and reduce camera shake. This is crucial for getting sharp eclipse images.
- Solar Filter: Never photograph the Sun without a solar filter specifically designed for photography. Solar filters protect your camera sensor and your eyes from harmful solar radiation. Use a solar filter that fits over

your camera lens.
- Remote Shutter Release: To minimize camera shake during exposure, use a remote shutter release or cable release. You can use your camera's self-timer function if you don't have one.

Camera Settings for Eclipse Photography:

- Manual Mode: Set your camera to manual (M) mode to fully control the exposure settings.
- ISO: Set the ISO to its lowest value (usually ISO 100 or 200) to minimize digital noise in your images.
- Aperture: Use a small aperture (high f-number, e.g., f/8 to f/16) to maximize depth of field and maintain sharpness.
- Shutter Speed: For partial eclipse phases, use a fast shutter speed (1/1000s or faster) with the solar filter in place to avoid overexposure.
- Focusing: Set your lens to manual focus and use Live View mode to zoom in on the Sun. Focus manually until the Sun appears sharp and clear.

Step-by-Step Guide to Your First Eclipse Photo:

- Preparation: Arrive at your chosen location well in advance. Set up your camera on a tripod, attach the solar filter to your lens, and ensure everything is stable. Have extra charged batteries on hand and your specific spare empty SD Card.
- Test Shots: Before the eclipse begins, take some test shots of the Sun with the solar filter on. Adjust your exposure settings to get a well-exposed image of the Sun's disk.
- Partial Phase: As the eclipse begins, continue taking photos at regular intervals to document the progression of the partial phase. Adjust your shutter speed to maintain a proper exposure as the Sun's brightness changes.
- Totality (if applicable): If you are in the path of totality, remove the solar filter ONLY during the brief moments of totality. Capture the beautiful

corona and any surrounding phenomena. Use a fast shutter speed (1/1000s or faster) to prevent overexposure during totality.
- Safety: Always wear eclipse glasses without your camera's viewfinder when looking at the Sun. Avoid looking at the Sun through your camera's optical viewfinder, as this can damage your eyes.
- Review and Adjust: Periodically review your images on the camera's LCD screen to ensure you're capturing the eclipse as desired. Make exposure adjustments as needed.
- Post-Processing: After the eclipse, transfer your images to a computer and use photo editing software to enhance and adjust the final images if necessary.

Remember that photographing a solar eclipse requires practice and patience. It's a challenging but rewarding endeavor, and with the right equipment and settings, you can capture stunning eclipse photos.

2. Advanced Techniques:

Long Exposure and HDR Techniques:

Long Exposure: Long exposure photography involves using slow shutter speeds (often several seconds or more) to capture multiple phases of an eclipse in a single image.

- To capture the entire eclipse progression in one photo, you'll need a stable tripod and a remote shutter release to minimize camera shake.
- Start with a solar filter during the partial phases and gradually increase the exposure time as the eclipse progresses.
- Combine multiple long-exposure images of different eclipse phases to create a composite image showcasing the entire event, from partial to totality.

HDR (High Dynamic Range): HDR photography involves capturing multiple shots at different exposures (bracketing) and then combining them to create an image with a broader dynamic range.

- Bracket your shots during the eclipse, ensuring you have well-exposed images of the Sun's disk, inner corona, and outer corona.

Use software like Adobe Photoshop or specialized HDR software to merge the bracketed images into one high-dynamic-range image that reveals details in both the bright and dark areas of the eclipse.

Capturing the Corona and Baily's Beads:

Capturing the Corona: To capture the ethereal solar corona during totality, remove your solar filter only during the brief period when the Moon completely covers the Sun. Use a fast shutter speed (1/1000s or faster) to prevent overexposure of the corona's delicate structures. Bracket your shots to capture the full range of corona brightness as it extends far from the Sun's surface.

Baily's Beads: Baily's Beads are bright spots of sunlight that appear just before and after totality when the last rays of sunlight pass through the Moon's valleys and create bead-like effects. To capture Baily's Beads, you need to take a rapid sequence of photos (burst mode) with a high shutter speed during the moments before and after totality. Combine these images into a single composite to showcase the beads forming and disappearing sequence.

Real-Life Examples of Stunning Eclipse Photos:

CHAPTER 5: ECLIPSE PHOTOGRAPHY FOR BEGINNERS

Totality and Coronal Detail: This stunning photo from the 2017 eclipse showcases both the intricate detail of the solar corona and the beauty of totality.

Diamond Ring Effect: The Diamond Ring Effect is a captivating moment just before or after totality when a single bead of sunlight remains, resembling a diamond ring. This image beautifully captures that phenomenon.

Totality and Landscape: These photographs capture the 2017 eclipse along with the landscape, providing a sense of the event's grandeur.

Advanced eclipse photography techniques require careful planning, practice, and patience. These techniques allow you to capture the eclipse and its stunning and ephemeral details, making for truly breathtaking and memorable images.

3. Post-Processing and Sharing:

Basic Editing Tips for Impact:

- Color Correction: Adjust the color balance to make sure the Sun appears true to life. Eclipse photos often enhance the orange hues of the solar corona.
- Contrast and Sharpness: Enhance contrast and sharpness to reveal details in the solar corona and the Moon's surface. Be cautious not to overdo it, as excessive processing can make the image look unnatural.
- Noise Reduction: Use noise reduction techniques to minimize any digital noise that may have been introduced during long exposures. Be mindful not to lose too much detail in the process.
- Cropping: Crop your photos to focus on the most captivating part of the image, such as the corona or a particularly stunning Baily's Beads moment.
- Composite Images: Consider creating composite images that blend different exposures to capture both the Sun's surface and the corona in a single shot. This can result in striking visuals.

Platforms for Sharing Your Work:

- Social media: Sharing eclipse photos is common on sites like Facebook, Instagram, and Twitter. Use pertinent hashtags, such #EclipsePhotography or #Astronomy, to connect with a larger audience.
- Photography Communities: You can display your eclipse photos on photography-focused websites such as 500px and Flickr. Additionally, you can join forums or groups dedicated to eclipses.
- Join astronomy discussion boards on sites such as Cloudy Nights or Stargazers Lounge. These groups frequently discuss and evaluate astrophotography projects.
- Photography Contests: Look for photography competitions that have to do with astronomy or eclipses. Gaining recognition or winning one of these competitions can provide you with a lot of exposure.

Narratives of Pictures That Got Viral:

- Diamond Ring Effect: Because of their breathtaking and dramatic look, pictures taken at or just after totality that capture the Diamond Ring Effect frequently become viral. Due to its transient nature, photographs of this phenomenon are greatly sought after.
- Original Compositions: Eclipse photographs with a standout foreground, such as famous sites, animals, or people in silhouette, may attract notice for their imaginative arrangement.
- Scientific Interest: Pictures that document certain scientific events during an eclipse, such as the "Bailey's Beads Sequence," can spark curiosity among scientists and the general public.
- Impact on Emotion: Photographs of eclipses that arouse powerful feelings in viewers, whether via narrative or creative expression, are more likely to do so. A picture that captures the wonder of the eclipse or conveys a personal tale may go viral.

Remember that while it's exciting to have your photos go viral, the most important aspect of sharing your eclipse images is the opportunity to inspire and educate others about the wonders of the cosmos. Whether your images are seen by a few or by millions, each one contributes to the collective appreciation of these celestial events.

Now that you're ready to photograph the eclipse let's explore how this cosmic event can inspire your creative expression.

Chapter 6: Creative Expressions of the Eclipse

The artistic impact of a solar eclipse on art is profound and has inspired countless artists over the centuries. Eclipses often serve as a source of creative inspiration due to their dramatic and otherworldly nature.

1. Eclipse Art and Literature:

Eclipses have inspired artists and writers for centuries, resulting in a rich tapestry of eclipse-related art and literature. Writing poetry or short stories inspired by eclipses can be a powerful way to convey emotions and explore themes of darkness and light, mystery, and transformation. Artists can experiment with different styles and techniques to create eclipse-themed paintings or visual art that reflect their personal interpretations of this celestial event. Here are some examples and insights:

"Eclipse Over America" by Howard Russell Butler: This famous painting, created by American artist Howard Russell Butler, depicts the total solar eclipse of 1918. The artwork is known for accurately portraying the eclipse's scientific aspects.

"Eclipse of the Sun" by Edvard Munch: This lithograph by the renowned

Norwegian artist Edvard Munch captures the eerie atmosphere of a solar eclipse. It is part of Munch's exploration of themes related to anxiety and existentialism.

"A Connecticut Yankee in King Arthur's Court" by Mark Twain: Mark Twain's novel features an eclipse as a central plot point. The protagonist, transported back in time to King Arthur's era, uses his knowledge of an upcoming eclipse to manipulate events and establish his authority.

"Eclipse" by John Banville: In John Banville's novel "Eclipse," the eclipse serves as a metaphor for the protagonist's internal struggles and existential ponderings.

2. Music and the Eclipse:

Eclipses have influenced performances and music throughout history. Musicians and performers can write songs, orchestrations, and theatrical productions to depict an eclipse's drama and emotional impact. Examples of the composition process and music with eclipse themes follows:

- "Eclipse" by Pink Floyd: "Eclipse" is a tune from the British rock band Pink Floyd's album "The Dark Side of the Moon." This song, which is frequently connected to eclipses, is a musical interpretation of several of the album's themes, such as the passing of time and existential reflection.
- Bonnie Tyler's "Total Eclipse of the Heart": Besides being one of the most well-known songs with an eclipse subject, this classic song is a potent ballad about strong feelings and experiences. Eclipse imagery is even featured in the song video.
- Eclipse Concerts and Events: Occasionally, orchestras and musicians plan concerts and other events with an eclipse theme to take place in tandem

CHAPTER 6: CREATIVE EXPRESSIONS OF THE ECLIPSE

with solar eclipses. These musical presentations frequently try to convey through song the wonder and drama of the cosmic event.

Musicians' Individual Narratives Motivated by Solar Eclipses:

- Gustav Holst's orchestral suite "The Planets" contains a movement titled "Mars, the Bringer of War," which might convey a sense of celestial drama similar to an eclipse, even if it was not directly inspired by one. Holst had a strong passion for astronomy and astrology.
- Solar Eclipse Musical Works: Several modern composers have created works of music that are primarily motivated by solar eclipses. Take into consideration listening to instrumental and classical pieces that generate a sense of cosmic awe and drama. These songs seek to express eclipses' distinct emotional and visual features through music. This theme is represented in works by composers such as Sergei Rachmaninoff, Gustav Holst, and Richard Strauss.
- Listen to and explore modern and ambient music that reflects an eclipse's contemplative and ethereal qualities. Ludovico Einaudi and Brian Eno are two musicians whose music might enhance an eclipse viewing experience.
- Soundscapes and Sounds of Nature: Incorporate ambient noises from nature, like soft waves or birdsong, to create a tranquil eclipse environment. For a distinctive soundtrack, combine these noises with ambient pieces.
- Personalized Playlists: Assemble a playlist of your preferred songs that express the sentiments and emotions you will be experiencing throughout the eclipse. Selecting music for an occasion is a highly personal decision, and your playlist can showcase your unique relationship to it.

You can add an artistic and emotional element to the visual spectacle of the eclipse by making your own soundtrack or listening to music with eclipse themes. Music, whether it be ethereal classical pieces or emotional rock ballads, helps us relate to eclipses more deeply.

3. Eclipse in Digital Media:

Eclipses provide for an intriguing subject for digital artists and photographers. One way to express oneself artistically is to photograph an eclipse or use an eclipse as inspiration for creative digital art.

Documentary Films: Solar eclipses are a common subject for documentary films. These movies depict eclipses' emotional impact, cultural relevance, and scientific importance. Interviews with astronomers and eclipse chasers might also be included. Examples of some of these films are below.

- "Eclipse: The Science and History of Nature's Most Spectacular Phenomenon" (2017): This documentary, produced by NOVA and PBS, explores the science, history, and cultural significance of solar eclipses. It features interviews with scientists and stunning footage of eclipses from around the world.
- "Eclipse Over America" (2017): Produced by the Science Channel, this documentary follows the path of the total solar eclipse that crossed the United States on August 21, 2017. It captures the excitement and scientific observations surrounding this historic event.
- "Chasing Shadows: The Life and Death of Peter Short" (2016): This documentary tells the story of eclipse chaser Peter Short and his lifelong quest to witness and photograph solar eclipses. It offers a unique perspective on the passion of eclipse enthusiasts.
- "Eclipse Across America" (2017): Produced by National Geographic, this documentary covers the 2017 total solar eclipse as it made its way across the United States. It includes interviews with experts and showcases the experiences of people along the eclipse path.
- "Eclipse: Total Darkness of the Sun" (2013): This documentary provides a comprehensive look at solar eclipses, delving into the science, history, and cultural impact of these celestial events. It includes stunning visual footage and expert interviews.
- "In the Shadow of the Moon" (2007): While not exclusively focused

CHAPTER 6: CREATIVE EXPRESSIONS OF THE ECLIPSE

on eclipses, this documentary explores the experiences of the Apollo astronauts who journeyed to the Moon and witnessed solar eclipses from space. It offers a unique perspective on eclipses as seen from lunar orbit.

- "Eclipse: Beyond the Discovery" (2000): Produced by the Discovery Channel, this documentary delves into the science and history of solar eclipses, featuring expert commentary and footage of various eclipse events.
- Eclipse Live Streams: Observatories, space agencies, and amateur astronomers frequently offer live web feeds of significant eclipses. Even though they cannot be on the path of totality, individuals from around the world may see the eclipse in real time, thanks to these live feeds.
- Eclipse Chaser Vlogs: People who follow eclipses often use video diaries and vlogs to record their experiences. These videos offer an intimate and realistic look at what it's like to travel to witness an eclipse.

Using Digital Channels to Disseminate Eclipse Stories:

- Social media: Sharing eclipse experiences is common on sites like Instagram, Twitter, Facebook, and TikTok. To establish connections with people who are also witnessing the eclipse, using pertinent hashtags and geotags.
- YouTube: You may upload videos about eclipses on YouTube, which is a great platform. You can post eclipse time-lapse films, eclipse photography guides, or personal vlogs documenting your eclipse experience.
- Blogs and Websites: To share your eclipse-related images, films, and tales, start a blog or website. This makes it possible to relate stories in greater detail and can be a useful tool for those who are curious about eclipses.
- Online Forums and Groups: Participate in online forums and groups pertaining to eclipses to exchange experiences and gain knowledge from others. Discussions on eclipses are ongoing on websites such as Cloudy Nights and Stargazers Lounge.

Taking Part in Eclipse Online Communities:

- Join communities and forums devoted to eclipse chasing by visiting eclipse chaser forums. There are plenty of seasoned eclipse chasers in these areas who can offer insightful counsel, pointers, and details about impending eclipses.
- Join social media communities about eclipses to gain insight into the phenomenon. These groups are excellent for exchanging content, connecting with other enthusiasts, and learning about eclipse-related news.
- Virtual Star Parties: Some online communities host virtual star parties during eclipses so that participants worldwide can view the eclipse together via live streaming and discuss their observations in real time.
- Astronomy Apps and Websites: Keep track of impending eclipses and receive notifications on their location and date by using real-time apps and websites. Interactive maps and tools for organizing eclipse viewing are frequently available on these sites.
- Participating in virtual eclipse communities can augment your eclipse encounter by facilitating connections with like-minded users, enabling you to exchange enthusiasm and gain knowledge from others. It's a means of extending the eclipse experience past the actual moment and fostering a feeling of worldwide eclipse solidarity.

Eclipses don't just inspire art; they also bring people together. Let's explore the social and communal aspects next.

Chapter 7: Environmental and Wildlife Responses to Eclipses

During the total solar eclipse of August 11, 1999, at the Welsh Mountain Zoo in Colwyn Bay, Wales, zookeepers and visitors noticed that the chimpanzees began to display unusual behavior. Instead of going about their everyday activities, the chimps gathered together and became notably quieter. They appeared to be observing the sky with a sense of awe and curiosity. Some even climbed to the top of their climbing frame to get a better view of the eclipse.

This unexpected behavior of the chimpanzees, which seemed to indicate their awareness of the celestial event, fascinated both scientists and zoo visitors. While the exact reason for their reaction remains a subject of speculation, it's a remarkable example of how even animals can respond to a solar eclipse's unique and awe-inspiring phenomenon.

1. Impact on Wildlife:

Here are some details on how animal behavior varies during eclipses, based on observations, studies, and firsthand accounts:

Animal Behavior Shifts During Eclipses:

Certain animals, particularly those that are typically active during the day, display nocturnal behavior when there is a solar eclipse. This includes diurnal mammals becoming more active at night and birds stopping their singing and returning to their nests.

- Confusion: Because of the abrupt shift in light levels during an eclipse, animals may experience confusion or disorientation. This may impact their eating and mating habits. Dogs, for example, may become anxious due to the unusual change in lighting conditions, while cats may become more active or seek shelter. In the wild, animals such as deer, squirrels, and rodents may exhibit changes in behavior during an eclipse. They might seek shelter, become more active, or display signs of confusion as they respond to the sudden darkness.
- Silence: During an eclipse, many birds and insects become noticeably quieter. It is common to refer to this abrupt quiet as the "eclipse hush." This behavior may be linked to the sudden decrease in light, which signals to birds that it's time to roost or seek shelter. Some researchers also believe that birds perceive the eclipse as an approaching nighttime, prompting them to prepare for rest. The brief darkness may have an impact on birds that depend on visual cues for navigation.
- Roosting and Nesting: When nightfall comes, birds may return to their roosts or nests. On the other hand, nocturnal animals, such as bats and owls, may become more active during a solar eclipse. The sudden darkness mimics the onset of nighttime, prompting these creatures to hunt or become more alert. This behavior is often seen in animals that rely on low-light conditions for their activities.
- Diurnal Insects: Insects like bees and butterflies may respond to an eclipse by ceasing their flight and seeking shelter. The decrease in light can disrupt their foraging patterns, leading them to rest until conditions return to normal.
- Aquatic Life: Some aquatic animals, such as fish and amphibians, may

become more active during an eclipse. The sudden drop in light can mimic the onset of evening, prompting these animals to engage in activities they typically do at dusk or night.

Investigations and Findings:

- Scientific Research: Studies have been done to see how animal behavior changes during eclipses. This research frequently shows animal movements, bird melodies, and other behavioral patterns.
- Observations at Zoos and Wildlife Parks: During eclipses, animal behavior is observed in carefully monitored settings like zoos and wildlife parks. Scientists and zookeepers have recorded how caged animals respond to solar and lunar eclipses.
- Jane Goodall: A renowned primatologist, Jane Goodall has seen eclipses while chimpanzees are in the wild. She pointed out that the chimpanzee may exhibit such characteristics as congregating and becoming quieter.
- Researchers studying wildlife: Numerous biologists and wildlife specialists have given their own accounts of animal behavior during eclipses. These reports advance our knowledge of the responses of different animals to this astronomical event.

All things considered, research is still ongoing to determine the precise mechanisms underlying the behavioral changes observed in animals during eclipses. Eclipses offer a rare chance to witness how various species respond to abrupt shifts in the natural light, providing essential insights into the flexibility and sensitivity of the animal kingdom to environmental cues.

2. Eclipse Effects on the Environment:

The weather, atmosphere, and environment can all be fascinatingly affected by eclipses. Below is more information on each of these subjects.

Brief Variations in the Atmosphere and Weather:

- Wind Patterns: Local wind patterns can be affected by temperature variations brought on by eclipses. Warm air rises, and cool air falls, which may cause changes in the direction and strength of the wind.
- Variations in Atmospheric Pressure: The dynamics and structure of the ionosphere, Earth's upper atmosphere, can be significantly impacted by solar eclipses. This is mainly because, during the eclipse, a dramatic drop in solar radiation reached Earth's atmosphere. Changes to the ionosphere can impact radio communications and navigation systems because it is made up of charged particles called electrons and ions responsible for reflecting and refracting radio signals. Solar eclipses can affect the ionospheric dynamics and structure in the following ways:
- Changes in Ionization: Ionization is the process by which an atom or molecule becomes charged when it receives or loses electrons, and it is mainly caused by solar radiation in the ionosphere. Ionization decreases during a solar eclipse due to a decrease in solar radiation, especially in the ionosphere area between 37 and 56 miles above the ground.
- Temperature Changes: During a solar eclipse, temperature drop is one of the most apparent impacts. There may be a brief cooling effect due to the temperature dropping dramatically as the Sun's direct energy is blocked. The reduction in solar radiation during an eclipse can also result in a short dip in the density of electrons in the ionosphere due to this decrease in ionization. Ionospheric densities and elevations may alter as a result of this cooling's impact on the ionosphere's temperature structure.
- Variations in Electron Density: The ionosphere's electron density may drop due to an eclipse's cooling and decreased ionization. This may have an impact on radio wave propagation, especially in the high frequency (HF) region, where long-distance communication relies on ionospheric reflection.
- Ionospheric Anomalies: Ionospheric anomalies, such as ionospheric holes or depletions, can emerge as a result of the abrupt changes in ionospheric conditions that occur during an eclipse. These anomalies

CHAPTER 7: ENVIRONMENTAL AND WILDLIFE RESPONSES TO ECLIPSES

may interfere with GPS and radio signals, making it difficult to navigate and communicate in the impacted areas.

- Propagation Effects: During an eclipse, the ionosphere's changing characteristics might affect how radio waves travel across it. This may result in signal refraction, absorption, and fading, which could impact satellite and shortwave communication systems.

Solar eclipses can significantly affect the ionosphere; however, these effects are usually transient and limited to the area where the eclipse occurs. Once the eclipse event ends, solar radiation levels return to normal and the ionosphere to its original state.

Examples of Particular Environmental Phenomena:

- Shadow Bands: Seldom seen and difficult to see, shadow bands are seen both before and after totality. The final rays of sunlight traveling through gaps in the Moon's valleys generate atmospheric turbulence, which is seen as rippling, wavy lines of alternating light and dark on the Earth.
- Animal Reactions: During eclipses, certain animals display unusual behaviors. For instance, nocturnal creatures may become active, and birds may return to their roosts. These responses offer case studies of how environmental cues impact wildlife.
- Eclipse-Induced Winds: Researchers have observed abrupt changes in wind patterns during total solar eclipses. When the temperature drops and localized wind disruptions result, a phenomenon called the "Eclipse Wind" happens.

Eclipses' Ecological Significance:

- Research on Ecology: Ecological research can benefit from eclipses. Researchers can examine the reactions of ecosystems and animals to sudden variations in temperature and light. Understanding plant physiology, animal behavior, and ecological dynamics can all be gained from this

research.
- Natural Clocks: Eclipses are natural experiments that throw off the regular daily cycles of plants and animals. It can be revealed how crucial natural cycles of light and temperature are to the survival of certain species by studying how they adjust to these disturbances.
- Awareness of Conservation: Since eclipses frequently draw large crowds of people, they offer an extraordinary chance to spread knowledge about environmental preservation. Eclipses can serve as educational opportunities for educators and conservationists to highlight how intertwined Earth's systems are.
- Historical Records: Documents detailing eclipse-related phenomena from the past, such as observations of the weather or animal behavior, might shed light on how earlier societies experienced and recorded similar occurrences. These documents advance our knowledge of ecological history.

Eclipses provide a dynamic, natural laboratory for researching how the environment, living things, and celestial events interact. Through the study of distinct environmental phenomena, the examination of transitory variations in weather and atmosphere, and the recognition of the ecological significance of eclipses, scientists and enthusiasts can get a more profound understanding of the complex interplay between Earth and the cosmos.

3. Human Impact and Sustainability:

The Environmental Footprint of Eclipse Tourists: It's essential to be conscious of eclipse tourists' potential sound and adverse environmental effects.

- Positive Impact: By drawing tourists, eclipse events can strengthen local economies by raising money for local companies and communities. This economic boom can fund infrastructure upgrades and conservation

CHAPTER 7: ENVIRONMENTAL AND WILDLIFE RESPONSES TO ECLIPSES

initiatives.
- Adverse Effect: Large populations of people traveling to eclipse viewing areas may stress local ecosystems and resources. Pollution and habitat disruption can result from increased energy use, waste production, and emissions from transportation.
- Light Pollution: Using artificial illumination during eclipse events can negatively impact local fauna and disturb the ecosystem, particularly in places where light pollution laws are strictly enforced.

Eco-Friendly Methods for Eclipse Spectators:

The following sustainable behaviors can be adopted by eclipse tourists to reduce the impact on the environment:

- Carpooling and Public transit: When traveling to eclipse viewing locations, cut down on carbon emissions by carpooling or taking public transit. Air pollution and traffic congestion are reduced as a result.
- Leave No Trace: Pack out all your garbage and rubbish per the Leave No Trace philosophy. Reduce your environmental impact, recycle whenever you can, and ethically dispose of trash.
- Select Eco-Friendly Lodging: Choose lodgings that emphasize sustainability, including eco-lodges or hotels that have earned green certifications. Give your support to companies that care about the environment.
- Employ Renewable Energy: When setting up your equipment, consider employing solar energy or other renewable energy sources for charging and lighting.
- Honor the Wildlife: Consider the ecosystems and wildlife in your area. Tales of Eco-friendly Eclipse Experiences: There are heartwarming accounts of people and towns adopting eco-friendly measures during eclipse events. Try not to disturb animals and their habitats.
- Community-Led Cleanups: Some communities plan post-event cleanups to keep the area spotless after the eclipse. To reduce their environmental impact, volunteers gather to collect trash.

- Eco-aware Travel Agencies: Tour operators who specialize in eclipse vacations often prioritize Eco-friendly techniques. They select lodging and modes of transportation following the principles of sustainable tourism.
- Initiatives for Education: Eclipses might provide chances for teaching about the environment. The organizers of eclipses occasionally collaborate with conservation organizations to spread awareness about preserving local ecosystems and fauna.
- Solar-Powered Events: A few eclipse events provide attendees with a demonstration of renewable energy solutions by setting up solar-powered stages and lights.

In conclusion, there may be an environmental cost associated with eclipse tourism; however, these costs can be offset by using sustainable behaviors. People who watch eclipses must practice environmental consciousness, patronize eco-friendly companies, and preserve the beauty of the natural world. Narratives of environmentally conscious eclipse-viewing gatherings serve to emphasize the constructive roles that people and groups can play in conserving the environment while taking in these beautiful celestial spectacles.

Understanding how eclipses affect our world leads us to ponder their deeper meanings. Let's explore the philosophical and existential aspects next.

Chapter 8: The Philosophical and Existential Dimensions of Eclipses

"The cosmos is within us. We are made of star-stuff. We are a way for the universe to know itself." - Carl Sagan

This quote by the renowned astrophysicist and science communicator Carl Sagan reflects the idea that the elements that make up our bodies and the universe itself were forged in the hearts of stars. It emphasizes the profound connection between humanity and the cosmos, inviting contemplation about our place in the vastness of the universe and the mysteries it holds.

1. Eclipses and Human Consciousness:

The Eclipse as a Metaphor for Life and Change: Eclipses have long been utilized in literature and other artistic mediums to symbolize life and change because of their striking flashes of light and blackness. They stand for the interplay of light and shadow, the cyclical aspect of life, and the certainty of change. Here are some things to think about:

- Cycles of Life: Since eclipses happen on a regular basis, they might be interpreted as a metaphor for life's cycles. Similar to how the Moon alter-

nates between light and dark, our lives go through stages of development, difficulty, and rebirth.
- Transformation: Life's unforeseen turns are analogous to the abrupt blackness of a total solar eclipse, which is followed by the appearance of the Sun's corona. It serves as a reminder that change and rebirth are possible even in the depths of darkness. Eclipses are a symbol of the delicate equilibrium that exists between conflicting forces. Finding balance in the face of change and uncertainty is a typical struggle in life. We can be motivated to pursue these challenges by our experiences viewing eclipses.
- Personal Narratives of Life-Changing Eclipse Experiences: There are a lot of first-hand stories from people who claim that eclipses have had a profound spiritual impact or changed their lives. These tales frequently center on the deep feelings and realizations brought about by seeing an eclipse. While some are motivated to make significant life changes, others experience a sense of oneness with the universe.
- Spiritual Awakening: During a total solar eclipse, several people claim to have had a strong sense of spiritual connection or awe. The union of heavenly bodies can evoke a feeling of unity with the universely bodies.
- Reevaluation of Priorities: Observing an eclipse's transitory nature might cause people to take stock of their own lives and reassess their priorities. It could act as a reminder of the transience of material worries.
- Motivation for Modification: Eclipse chasers talk a lot about how these heavenly phenomena are addicting. Some people have seen significant life shifts due to their love for chasing eclipses; they have chosen to work as astronomers or environmental conservationists.

Philosophical Views on Eclipses:

Throughout history, philosophers and intellectuals have proposed various explanations for eclipses. A few philosophical stances are as follows:

- Plato & metaphor: Plato described how people see reality by using the

metaphor of the cave. An eclipse might be viewed as a fleeting window of insight, a look beyond the shadows of daily life.
- Existential Reflection: According to existentialist philosophers such as Jean-Paul Sartre, the human state might be metaphorically represented by an eclipse. The inherent uncertainty and choices in life can be compared to the darkness and ambiguity of an eclipse.
- Harmony and oneness: The concept of cosmic harmony and oneness is emphasized in certain intellectual traditions. Eclipses can represent the interdependence of all things in the universe due to their celestial nature.

Eclipses have the ability to provoke in-depth philosophical reflection on the nature of change, life, and our place in the universe. They provide a chance for introspection and philosophical thought while serving as a reminder of the wonder and beauty of the cosmos.

2. Spiritual and Mystical Aspects

Eclipses in Spiritual Practices and Beliefs:

For centuries, eclipses have been spiritually significant to societies all across the planet. They are frequently seen as potent cosmic occurrences with the ability to affect human fate. Eclipses are incorporated into spiritual activities and beliefs in the following ways:

- Symbolism: Eclipses represent the interaction of light and dark, and many spiritual traditions see this as a metaphor for life's dualities, including good and evil, the material and spiritual worlds, and life and death.
- Purification: Eclipses are said to have purifying properties in some civilizations. For instance, during an eclipse, it's customary in Hinduism to bathe ritualistically in order to purge oneself of bad energy.
- Mythological Significance: Eclipses are associated with stories and folk-

lore in many cultures. According to Norse mythology from antiquity, eclipses symbolized the conflict between light and dark and happened when celestial wolves were eating the Sun or Moon.
- Astrology and Divination: Based on the astrological positions of an eclipse, astrologers and diviners may provide instruction or interpret eclipses as significant omens.

Narratives of Spiritual Enlightenments Occurring During Eclipses:

Numerous people have conveyed their profound spiritual epiphanies and insights during eclipses. Even though these experiences are quite personal, the following elements are frequently present:

- Sense of Unity: Some people talk about having a strong sense of interconnectedness with the cosmos or being part of a larger cosmic whole. Feelings of transcendence and spiritual awakening may result from this union.
- Transformation and Renewal: Eclipses are said to be periods of renewal and transformation. A few people have reported a spiritual rebirth or epiphanies on their life's purpose during an eclipse.
- Deep Reflection: A total solar eclipse's abrupt darkness can foster an environment of reflection. During this period, people could reflect on their values, beliefs, and decisions in life.

The Ethereal Charm of Heavenly Occurrences:

Eclipses and other celestial events are fascinating because of their mystique, uniqueness, and ability to astonish. Several facets of its ethereal appeal are as follows:

- Uncommon Events: Since eclipses are comparatively uncommon occurrences, seeing one can seem like a once-in-a-lifetime opportunity. Their uniqueness heightens their enigma.

- Cosmic connectedness: We are reminded of our connectedness to the wide cosmos by celestial events such as eclipses. They highlight the universe's majesty and beauty and encourage reflection on our place in it.
- Amazement and Wonder: During an eclipse, one is filled with amazement and wonder due to the fantastic sights and abrupt changes in the sky. For many, this emotional reaction might have profound spiritual implications.
- Cultural Significance: Eclipses have contributed significantly to the mystique of religious and cultural myths by playing pivotal roles. They are frequently interwoven with rituals and cultural identities.

In conclusion, eclipses are revered in spiritual practices and beliefs throughout the world. They can elicit intense spiritual experiences that give people a sense of oneness with the universe and a chance for reflection and rejuvenation. The enigmatical charm of astronomical occurrences such as eclipses resides in their capacity to evoke awe and reflection over the secrets of the cosmos.

3. Contemplating Our Place in the Universe:

The Humbling Effect of Witnessing an Eclipse:

Eclipses have had spiritual significance for societies worldwide for millennia. They are often perceived as powerful cosmic events capable of influencing human destiny. The following are some ways that eclipses are incorporated into spiritual practices and beliefs:

- Eclipses symbolize the interplay of light and shadow, and many spiritual traditions see this as a metaphor for the dualities that exist in existence, such as life and death, the material and spiritual realms, and good and evil.
- Purification: According to some cultures, eclipses have purifying effects. For example, ritualistic bathing is typical in Hinduism after an eclipse to

cleanse oneself of negative energies.
- Mythological Significance: Eclipses are connected to legends and stories in many cultures. According to ancient Norse mythology, Eclipses occurred when the Sun or Moon was being devoured by celestial wolves and represented the struggle between light and dark.
- Astrology and Divination: Astrologers and diviners can offer guidance or read eclipses as important omens based on the eclipse's astrological placements.

Spiritual Enlightenment Stories Taking Place During Eclipses: Many people have shared their profound spiritual realizations and insights during eclipses. These are very personal experiences; however, the following components are standard:

- Sense of Unity: Some individuals discuss feeling deeply a part of a greater cosmic whole or connected to the cosmos. This union may cause feelings of transcendence and spiritual awakening.
- Transformation and regeneration: According to popular belief, eclipses are times of transformation and regeneration. A few people have described experiencing a spiritual rebirth or epiphanies regarding their life's purpose during an eclipse.
- Deep Reflection: The sudden darkness of a total solar eclipse can create a contemplative atmosphere. People had the opportunity to reflect on their life decisions, values, and beliefs at this time.

The Airy Allure of Celestial Events:

Because of their secrecy, distinctiveness, and capacity to amaze, eclipses and other cosmic events are interesting. Its ethereal appeal has the following aspects:

- Rare Occurrences: Seeing an eclipse can seem like a once-in-a-lifetime event because they are relatively rare occurrences. Their distinctiveness

adds to their mystique.
- Cosmic connectedness: A celestial event like an eclipse reminds us of our connection to the vast cosmos. They draw attention to the majesty and beauty of the universe and inspire contemplation about our role in it.
- Shock and Wonder: The breathtaking visuals and sudden shifts in the sky during an eclipse leave one feeling amazed and in awe. This emotional response could have significant spiritual ramifications for a lot of people.
- Cultural Significance: Because eclipses are so important, they have greatly added to the mystique surrounding religious and cultural mythology. They are often entwined with cultural identities and traditions.

In conclusion, eclipses are highly respected in global spiritual practices and beliefs. They have the power to provoke profound spiritual experiences that allow people a chance for introspection and renewal, in addition to a feeling of unity with the cosmos. The mysterious allure of celestial events such as eclipses lies in their ability to provoke wonder and contemplation about the mysteries of the universe.

As we approach the end of our journey, let's look forward to future eclipses and how you can continue your celestial adventures.

Chapter 9: Looking Ahead - Future Eclipses and Continuing Your Journey

1. Upcoming Eclipses

Calendar of future solar eclipses

DATE	SOLAR ECLIPSE TYPE	LOCATIONS OF GREATEST SOLAR OBSCURATION
04/08/2024	Total	Northern Mexico, Texas to New England, north Atlantic
10/02/2024	Annular	Southeast Pacific, far southern South America
03/29/2025	Partial	Europe, Asia, Africa, North America, South America, Atlantic Ocean, Arctic Ocean
09/21/2025	Partial	Australia, Antarctica, Pacific Ocean, Atlantic Ocean
02/17/2026	Annular	Antarctica
08/12/2026	Total	Arctic, eastern Greenland, Iceland, northern Spain
02/06/2027	Annular	South Pacific, southern Chile, southern Argentina, south Atlantic
08/02/2027	Total	Central Atlantic, Mediterranean region, Egypt, Red Sea area

CHAPTER 9: LOOKING AHEAD - FUTURE ECLIPSES AND CONTINUING...

Calendar of future lunar eclipses

DATE	LUNAR ECLIPSE TYPE	LOCATIONS OF GREATEST LUNAR OBSCURATION
Mar 13-14, 2025	Total	Europe, Much of Asia, Much of Australia, Much of Africa, North America, South America, Pacific, Atlantic, Arctic, Antarctica.
Sep 7-8, 2025	Total	East in Europe, Asia, Australia, North America, South America, Pacific, Atlantic, Indian Ocean, Arctic, Antarctica.
Mar 2-3, 2026	Total	East in Europe, Asia, Australia, North America, South America, Pacific, Atlantic, Indian Ocean, Arctic, Antarctica.
Dec 31-Jan 1, 2028	Total	Europe, Asia, Australia, Africa, North/West North America, Pacific, Atlantic, Indian Ocean, Arctic.
Jun 25-26, 2029	Total	Europe, West in Asia, Africa, North America, South America, Pacific, Atlantic, Indian Ocean, Antarctica.
Dec 20-21, 2029	Total	Europe, Asia, North/West Australia, Africa, North America, South America, Pacific, Atlantic, Indian Ocean, Arctic.
Apr 25-26, 2032	Total	South/East Europe, Asia, Australia, Much of Africa, Much of North America, Pacific, Atlantic, Indian Ocean, Antarctica.
Oct 19-19, 2032	Total	Europe, Asia, Australia, Africa, Much of North America, North/East South America, Pacific, Atlantic, Indian Ocean, Arctic, Antarctica.

After the total solar eclipse on April 8, 2024, the next total solar eclipse that can be seen from the contiguous United States will be on Aug. 23, 2044.

2. Resources for Further Learning in Astronomy

- Online Courses and Tutorials: There are numerous online platforms and websites that offer free or paid courses and tutorials in astronomy. Websites like Coursera, edX, and Khan Academy provide a wide range of astronomy courses suitable for beginners and advanced learners.
- Astronomy Books: Consider reading astronomy books written for general audiences. Some recommended titles include "Cosmos" by Carl Sagan, "Astrophysics for People in a Hurry" by Neil deGrasse Tyson, and "The Elegant Universe" by Brian Greene.
- Astronomy Organizations: Joining astronomy clubs or organizations can

provide valuable resources for learning and networking. The Astronomical Society of the Pacific (ASP) and the American Astronomical Society (AAS) are excellent examples.

- Planetariums and Observatories: Many cities have planetariums and observatories that offer public programs, lectures, and telescope viewing sessions. These can be a great way to learn and observe celestial objects.
- Astronomy Apps and Software: There are numerous astronomy apps and software available for smartphones and computers that help you explore the night sky, identify celestial objects, and learn more about them.

Opportunities for Amateur Astronomers

- Stargazing: One of the simplest and most rewarding activities for amateur astronomers is stargazing. You can do this from your backyard or join local astronomy clubs for group stargazing sessions.
- Telescope Ownership: Investing in a good-quality telescope lets you observe planets, stars, and deep-sky objects. Research different types of telescopes and choose one that suits your interests and budget.
- Astronomy Outreach: Many amateur astronomers engage in outreach activities, such as giving public talks, organizing star parties, and introducing astronomy to schools and communities.
- Astronomy Photography: Astrophotography is a popular hobby among amateur astronomers. You can capture stunning images of celestial objects using specialized cameras and equipment.
- Citizen Science Projects: Participate in citizen science projects like the "Zooniverse," where you can contribute to real scientific research by analyzing astronomical data.

Inspiring Narratives of Ongoing Celestial Exploration

- NASA Missions: Follow NASA's missions, such as the Mars rovers, the James Webb Space Telescope, and the upcoming Artemis program to return humans to the Moon. NASA's website and social media channels

CHAPTER 9: LOOKING AHEAD - FUTURE ECLIPSES AND CONTINUING...

provide regular updates.
- SpaceX and Commercial Space Ventures: Companies like SpaceX are pioneering the future of space exploration. Stay informed about their missions, including plans for Mars colonization.
- Hubble Space Telescope: Explore the incredible images and discoveries made by the Hubble Space Telescope. The Hubble website offers a wealth of information and stunning visuals.
- Space Agencies Worldwide: Besides NASA, keep an eye on the European Space Agency (ESA), the Russian Space Agency (Roscosmos), and other space agencies for their ongoing missions and discoveries.
- Astronomical Observatories: Learn about the work done by significant observatories like Chile's Very Large Telescope (VLT) and the Atacama Large Millimeter/submillimeter Array (ALMA).
- Space News Websites and Documentaries: Follow space news websites and watch documentaries like "The Farthest: Voyager in Space" and "Cosmos: A Spacetime Odyssey" for inspiring narratives of celestial exploration.

These resources provide you with ample opportunities to learn about astronomy and stay inspired by ongoing stellar exploration. Whether you're a beginner or a seasoned enthusiast, there's always something new and exciting to discover in the world of astronomy.

3. Staying Engaged with the Eclipse Community

Dedicated eclipse enthusiast clubs and organizations welcome members with a passion for eclipses. Joining such clubs can provide you with a sense of community, access to resources, and opportunities to connect with like-minded individuals who share your interest in eclipses.

Numerous online forums and discussion boards cater to eclipse enthusiasts. Websites like EclipseChasers.org and Cloudy Nights have sections where enthusiasts can share their experiences, exchange tips, and discuss upcoming eclipses. These forums are excellent platforms for asking questions, sharing observations, and learning from others.

Platforms like Facebook, Reddit, and Twitter host eclipse-related groups and communities. Joining these groups allows you to connect with enthusiasts worldwide, stay updated on eclipse news, and share your own experiences and photos.

Organizing or Participating in Local Events:

- Eclipse enthusiasts often organize or participate in local viewing parties during solar or lunar eclipses. These events can range from watch gatherings in parks to organized events at observatories or science centers. It's a great way to share the eclipse experience with others in your community.
- Some eclipse enthusiasts take their passion a step further by organizing educational events at schools, libraries, or community centers. They may give presentations about eclipses, set up telescopes for public viewing, and provide information on eclipse safety.
- Enthusiasts often plan trips to locations with upcoming total solar eclipses. This involves researching the best viewing spots, making travel arrangements, and sometimes even joining eclipse-themed tours. Traveling for eclipses can create unforgettable experiences and lasting memories.

The Enduring Bond of the Eclipse Enthusiast Community:

Eclipse enthusiasts share a deep passion for forming strong bonds. Eclipse events, especially total solar eclipses, are awe-inspiring and create lasting memories. The community is a hub for continual learning, support, encouragement, and strengthening ties. Enthusiasts document experiences to share, allowing everyone to relive eclipse beauty. Joining clubs, online forums, and

CHAPTER 9: LOOKING AHEAD - FUTURE ECLIPSES AND CONTINUING...

local events enriches the eclipse enthusiast experience, fostering connection and learning among like-minded individuals.

As we close this guide, let's reflect on our incredible journey, from the science to the sublime, and the lasting impact of the 2024 Solar Eclipse.

Chapter 10: Conclusion

In this book, we've embarked on a captivating journey into the mesmerizing world of solar eclipses, uncovering their enchanting allure and profound mysteries. Our exploration has focused on the celestial wonder itself and the upcoming solar eclipse on April 8th, 2024, a celestial event with unique attributes that promise to leave an indelible mark on all who witness it. Throughout these pages, we've delved into the intricate science and rich historical context surrounding this awe-inspiring cosmic phenomenon.

However, our voyage doesn't end here; it extends an invitation to further discovery and engagement within the eclipse enthusiast community. As you close this book, consider it a stepping stone to an ever-deepening connection with the world of eclipses. Beyond these chapters, you'll find a world of opportunities to continue your eclipse education, participate in thrilling eclipse-related events, and forge lasting connections with kindred spirits who share your fascination.

As you gaze at the skies and ponder the mysteries of the cosmos, let this book be your guiding star, illuminating a path toward ongoing wonder, exploration, and involvement in the endlessly fascinating realm of solar eclipses. With each eclipse, there's a chance for renewed awe and understanding, reminding us of the boundless wonders that the universe holds for those who dare to look up and explore its secrets.

Chapter 11: Resources

Resources

Bartels, M. (2024, January 18). Why 2024's total solar eclipse will be so special. Scientific American. https://www.scientificamerican.com/article/why-2024s-total-solar-eclipse-will-be-so-special/

Carter, J. (2024, January 1). Total solar eclipse April 2024: 10 of the biggest cities within in the path of totality. Space.com. https://www.space.com/total-solar-eclipse10-of-biggest-cities-in-path-of-totality-april-8-2024

Eclipse Glossary - NASA Science. (n.d.). https://science.nasa.gov/eclipses/glossary

Eclipses. (2017, August 10). HISTORY. https://www.history.com/topics/natural-disasters-and-environment/history-of-eclipses

Eclipses - NASA Science. (n.d.). https://science.nasa.gov/eclipses/

Five tips for photographing the annular solar eclipse on Oct. 14 - NASA Science. (n.d.). https://science.nasa.gov/solar-system/skywatching/five-tips-for-photographing-the-annular-solar-eclipse-on-oct-14/

Guide to the 2024 Total Solar eclipse | Exploratorium. (2023, November 21). Exploratorium. https://www.exploratorium.edu/eclipse/2024-total-solar-eclipse-guide

The impact of solar eclipses on the structure and dynamics of Earth's upper atmosphere - NASA Science. (n.d.). https://science.nasa.gov/solar-system/skywatching/eclipses/solar-eclipses/the-impact-of-solar-eclipses-on-the-structure-and-dynamics-of-earths-upper-atmosphere/

National Eclipse. (n.d.). Texas | April 8, 2024 - Total Solar Eclipse. https://nationaleclipse.com/states/2024-total-solar-eclipse-texas.html

National Solar Observatory. (2023, July 26). Eclipse Map - April 8, 2024 - NSO - National Solar Observatory. NSO - National Solar Observatory. https://nso.edu/for-public/eclipse-map-2024/

CHAPTER 11: RESOURCES

Solar and lunar eclipses worldwide – next 10 years. (n.d.-b). https://www.timeanddate.com/eclipse/list.html

Solar Eclipse Folklore, myths, and superstitions. (n.d.). Almanac.com. https://www.almanac.com/solar-eclipse-folklore-myths-and-superstitions

Svs. (2023, March 8). The 2023 and 2024 Solar Eclipses: Map and data. SVS. https://svs.gsfc.nasa.gov/5073/#media_group_312816

Texas Eclipse viewing information for the Great North American Eclipse of April 8, 2024 | eclipse2024.org. (n.d.). Eclipse2024.org. https://eclipse2024.org/eclipse_cities/states.php?state=Texas&country=USA

Total totality in tupper. (n.d.). Tupper Lake. https://www.tupperlake.com/story/2023/total-totality-in-tupper

What is the umbra? (n.d.). https://www.timeanddate.com/eclipse/umbra-shadow.html

Wikipedia contributors. (2024, January 25). Solar eclipse. Wikipedia. https://en.wikipedia.org/wiki/Solar_eclipse

Share the joy — Leave a Review!

Did you know that by doing something nice for someone else, you can feel super awesome? Guess what? You've got a chance to feel that way right now, and I'm here to tell you how.

I've got a little favor to ask you...

Could you help out a fellow space explorer, someone just like you, who is eager to learn about the solar eclipse but might not know where to start?

My goal is to make "The Complete Guide to the 2024 Solar Eclipse" a book that everyone who's excited about the eclipse can enjoy and learn from. But to do that, I need to reach as many future astronomers as possible.

This is where you come in!

SHARE THE JOY — LEAVE A REVIEW!

Believe it or not, your opinion matters a LOT. Yep, when people are looking for a good book, they definitely check out what others think about it. So, here's what I'm asking for the sake of a curious eclipse watcher you haven't met:

Could you take a moment to leave a review for this book? It doesn't cost a thing and takes just a little bit of your time, but your review could make a big difference. It might help...

...a family plan an unforgettable eclipse-watching adventure.

...a teacher find new ways to show the wonders of space to their students.

...a young stargazer learn something exciting.

...a community come together to watch the eclipse.

...someone's dream of witnessing an eclipse come true.

Feeling good already? You can make this happen by leaving a review. It's super easy!

Just scan this QR code to share your thoughts:

If the idea of helping out someone you haven't met makes you happy, then you're awesome, and I'm thrilled to have you as part of this journey. Together, we're going to make watching the 2024 eclipse an unforgettable experience. I've packed this book with tips and strategies just for you.

A huge thank you for your support and kindness. Let's get back to preparing for the eclipse!

- Your guide and friend,

D Walter Stuart

P.S. - Did you know? When you help someone out, they're likely to think you're pretty cool. If you know someone who's pumped about the eclipse, why not share this book with them? It might just make their day.

Made in the USA
Middletown, DE
25 March 2024